What Mad Pursuit

BOOKS IN THE ALFRED P. SLOAN FOUNDATION SERIES

Disturbing the Universe *by Freeman Dyson*
Advice to a Young Scientist *by Peter Medawar*
The Youngest Science *by Lewis Thomas*
Haphazard Reality *by Hendrik B. G. Casimir*
In Search of Mind *by Jerome Bruner*
A Slot Machine, a Broken Test Tube *by S. E. Luria*
Rabi: Scientist and Citizen *by John Rigden*
Alvarez: Adventures of a Physicist *by Luis W. Alvarez*
Making Weapons, Talking Peace *by Herbert F. York*
The Statue Within *by François Jacob*
In Praise of Imperfection *by Rita Levi-Montalcini*
Memoirs of an Unregulated Economist *by George J. Stigler*

THIS BOOK IS PUBLISHED AS PART OF AN ALFRED P. SLOAN FOUNDATION PROGRAM

What Mad Pursuit

A Personal View of Scientific Discovery

FRANCIS CRICK

Basic Books, Inc., Publishers
New York

Library of Congress Cataloging-in-Publication Data

Crick, Francis, 1916–
What mad pursuit.

(Alfred P. Sloan Foundation series)
Includes index.
1. Crick, Francis, 1916– . 2. Biologists—England—Biography. 3. Physicists—England—Biography. I. Title. II. Series.
QH31.C85A3 1988 574.19'1'0924 [B] 88–47693
ISBN 0–465–09137–7

Printed in the United States of America
Designed by Vincent Torre
88 89 90 91 RRD 9 8 7 6 5 4 3 2 1

Experience is the name everyone
gives to their mistakes.

—Oscar Wilde

Contents

PREFACE TO THE SERIES ix

ACKNOWLEDGMENTS xi

Introduction 3

1. Prologue: My Early Years 7
2. The Gossip Test 15
3. The Baffling Problem 24
4. Rocking the Boat 39
5. The α Helix 53
6. How to Live with a Golden Helix 62
7. Books and Movies About DNA 80
8. The Genetic Code 89
9. Fingerprinting Proteins 102
10. Theory in Molecular Biology 108
11. The Missing Messenger 116
12. Triplets 122
13. Conclusions 137
14. Epilogue: My Later Years 143

APPENDIX A: A Brief Outline of Classical Molecular Biology 164

APPENDIX B: The Genetic Code 169

INDEX 171

Illustrations follow page 112

Preface to the Series

THE ALFRED P. SLOAN FOUNDATION has for many years had an interest in encouraging public understanding of science. Science in this century has become a complex endeavor. Scientific statements may reflect many centuries of experimentation and theory, and are likely to be expressed in the language of advanced mathematics or in highly technical terms. As scientific knowledge expands, the goal of general public understanding of science becomes increasingly difficult to reach.

Yet an understanding of the scientific enterprise, as distinct from data, concepts, and theories, is certainly within the grasp of us all. It is an enterprise conducted by men and women who are stimulated by hopes and purposes that are universal, rewarded by occasional successes, and distressed by setbacks. Science is an enterprise with its own rules and customs, but an understanding of that enterprise is accessible, for it is quintessentially human. And an understanding of the enterprise inevitably brings with it insights into the nature of its products.

The Sloan Foundation expresses great appreciation to the advisory committee. Present members include the chairman, Simon Michael Bessie, Co-Publisher, Cornelia and Michael Bessie Books; Howard Hiatt, Professor, School of Medicine, Harvard University; Eric R. Kandel, University Professor, Columbia University College of Physicians and Surgeons, and Senior Investigator, Howard Hughes Medical Institute; Daniel Kevles, Professor of History, California Institute of Technology; Robert Merton, University Professor Emeritus, Columbia University; Paul Samuelson, Institute Professor of Economics, Massachusetts Institute of Technology; Robert Sinsheimer, Chancellor Emeritus, University of California, Santa Cruz; Steven Weinberg, Professor of Physics, University of Texas at Austin; and Stephen White, former Vice-President of the Alfred P. Sloan Foundation. Previous members of the committee were Daniel McFadden, Professor of Economics, and Philip Morrison, Professor of Physics, both of the Massachusetts Institute of Technology; Mark Kac (deceased), formerly Professor of Mathematics, University of Southern California; and Frederick E. Terman (deceased), formerly Provost Emeritus, Stanford University. The Sloan Foundation has been represented by Arthur L. Singer, Jr., Stephen White, Eric Wanner, and Sandra Panem. The first publisher of the program, Harper & Row, was represented by Edward L. Burlingame and Sallie Coolidge. This volume is the seventh to be published by Basic Books, represented by Martin Kessler and Richard Liebmann-Smith.

—ALBERT REES
President
Alfred P. Sloan Foundation
September 1988

Acknowledgments

THIS BOOK was started at the suggestion of the Sloan Foundation, for whose generous support I am most grateful. I was approached in 1978 by Stephen White, who persuaded me to sign the initial memorandum of agreement but I was very dilatory about beginning to write. I might have stayed in this state indefinitely but for Sandra Panem, who took over as book program director in 1986. She liked the idea of the book that was forming in my mind, and stimulated by her enthusiastic encouragement I produced a first draft. This was expanded and improved enormously as the result of her detailed comments, together with those of the Sloan Advisory Committee. I have also been helped by the comments of Martin Kessler, Richard Liebmann-Smith, and Paul Golob of Basic Books and by the copy editor, Debra Manette, who has improved my English in many places. I am also grateful to Ron Cape, Pat Churchland, Michael Crick, Odile Crick, V. S. (Rama) Ramachandran, Leslie Orgel, and Jim Watson, all of whom made helpful comments on one or another of the earlier drafts.

In writing the rest of the book, I have not made a deliberate attempt to acknowledge those who have been very close associates and have also influenced me strongly. While I shall not try to list here all my many friends and colleagues, there are three whom I must single out for special mention. The text does make clear how much I owe to Jim Watson. It does less than justice to my long and very fruitful association with Sydney Brenner. He was my closest associate for almost twenty years, and during much of that time we had long scientific discussions on almost every working day. His clarity, incisiveness, and fertile enthusiasm made him an ideal colleague. My third debt is to Georg Kreisel, the mathematical logician, whom I always address by his last name in spite of our having known each other for about forty-five years. When I met Kreisel I was a very sloppy thinker. His powerful, rigorous mind gently but steadily made my thinking more incisive and occasionally more precise. Quite a number of my mental mannerisms spring from him. Without these three friends my scientific career would have been very different.

My other major debt is to my family. Not only did they encourage me to become a scientist but they helped me financially. My parents made considerable sacrifices to enable me to go away to boarding school, especially during the Depression. My uncle Arthur Crick and his wife not only assisted me financially while I was a graduate student at University College but also gave me the money to buy our first house. My aunt Ethel, in addition to teaching me to read, helped financially when I first went to Cambridge after the war, as did my mother. They both helped also with the education of my son Michael. While I had very little money when I was young, I was secure in the knowledge that, thanks to my relatives, I would have enough to live on.

During most of the period covered by the main sections of this book I was employed in Cambridge by the British Medical Research Council. I am especially grateful to them, and in particular to Sir Harold Himsworth (then Secretary of the MRC) for providing such perfect working conditions there for me and my colleagues.

I should also record my gratitude to my present employer, The Salk Institute for Biological Studies, and in particular its president, Dr. Frederic de Hoffmann, for allowing me to work in such a delightful and stimulating atmosphere.

While writing this book I was mainly occupied in studying the brain. I thank the Kieckhefer Foundation, the System Development Foundation, and the Noble Foundation for their financial support of my efforts.

I thank the Editor of *Nature* for allowing me to quote at length from my article entitled, "The Double Helix: A Personal View," published on April 26, 1974; the New York Academy of Sciences for permission to quote extensively from an article of mine, "How to Live with a Golden Helix," which appeared in *The Sciences* in September 1979; Richard Dawkins and W. W. Norton and Company for permission to use several passages from his book, *The Blind Watchmaker,* published in 1986; V. S. (Rama) Ramachandran and Cambridge University Press for allowing me to quote a paragraph from his chapter "Interactions Between Motion, Depth,

Color, Form and Texture: The Utilitarian Theory of Perception," soon to appear in *Vision, Coding and Efficacy,* edited by Colin Blakemore; and Jamie Simon for doing the drawings.

Finally, my warmest thanks to my secretary, Betty ("Maria") Lang, who has coped splendidly with the many successive versions and all the tedious chores associated with producing a manuscript.

What Mad Pursuit

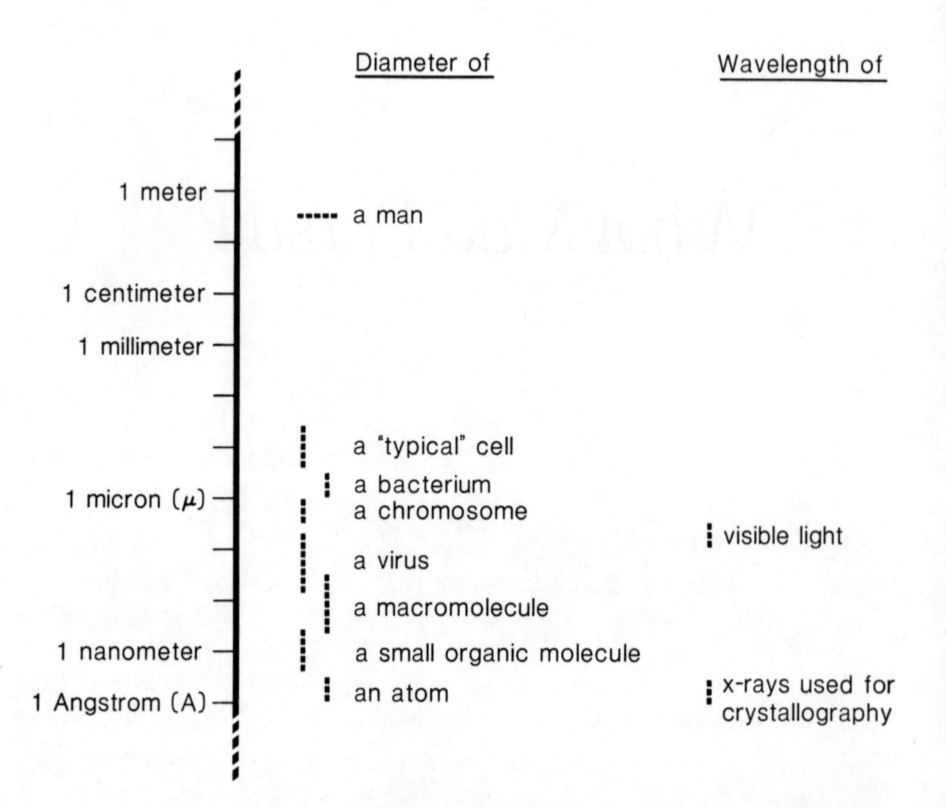

FRONTISPIECE

To show the approximate size of various objects, from molecules to man. Note that each step in the scale is a factor of ten.

Introduction

THE MAIN PURPOSE of this book is to set out some of my experiences before and during the classical period of molecular biology, which stretched from the discovery of the DNA double helix in 1953 till about 1966 when the genetic code—the dictionary relating the nucleic acid language to the protein language—was finally elucidated. As a preliminary I have put a short prologue that outlines a few details of my upbringing and education, including my early religious education, followed by an account of how I decided (after the Second World War) what branch of science to study, using the "gossip test" to help me. I have also included an epilogue, describing in outline what I have been doing since 1966.

There is an important difference between the scientific work described in the main body of the book and that touched on in the epilogue. In the former case we know with reasonable certainty what the correct answers are (the protein-folding problem is an exception). In the epilogue we do not yet know how things will turn out (the exception here is the double helix). For this reason many of my remarks in the epilogue are a matter of opinion. My comments in the main body of the book have somewhat more authority. One of the striking characteristics of modern science is that it often moves so fast that a research worker can see rather clearly whether his earlier ideas, or those of his contemporaries, were correct or incorrect. In the past, this opportunity did not arise so often. Nor does it today in slowly moving fields.

I have not tried to give an exhaustive account of what I did scientifically during those exciting years, let alone the large amount of work done by others. For example, I have said little or nothing about the ideas Jim Watson and I had about virus structure, nor about my collaboration with Alex Rich on a number of molecular structures. Instead I have included only those episodes that seem to me to have some general interest or to teach some general lesson about how research is done and what mistakes to avoid, especially those mistakes most relevant to biology. To do this I have to dwell somewhat more on errors than on successes.

In 1947, at the age of thirty-one, I went to Cambridge. After about two years working at the Strangeways Laboratory (a tissue-culture lab) I transferred to the Cavendish—the physics laboratory. There I became a graduate student again, trying to learn something about the three-dimensional structure of proteins by studying the X-ray diffraction patterns produced by protein crystals. It was then that I first learned how to go about doing research. It was during this period, while I was still a graduate student, that Jim Watson and I put forward the double-helical structure of DNA.

It has been difficult for me to write anything very new about the events leading up to the discovery of the double helix, since this has already been the subject of several books and movies. Rather than go over such familiar ground once again, I have found it better to comment on various aspects of the discovery and also on the recent BBC television movie *Life Story,* which deals with the discovery. In the same way I have not spelled out exactly how the genetic code was discovered—this is outlined in almost all modern textbooks. Instead I have dwelt mainly on the ups and downs of the theoretical approach, because I think few people realize exactly what a failure all this theoretical work on the genetic code turned out to be.

Since I am concerned more with ideas than with people, I have not included detailed character sketches of my friends and colleagues, mainly because I am reluctant to write candidly about close personal relationships with people still alive. In spite of this I have scattered through the text a number of small anecdotes, to give at least a few glimpses of what scientists are like, and to make for easier reading. Few people will willingly slog through an uninterrupted intellectual argument that lasts a whole book, unless they are acutely interested in the topic. In short, my main aim has been to put over a few ideas and insights in what I hope is an entertaining manner.

I have written both for my fellow scientists and for the general public, but I believe a layman can easily understand most of what I discuss. Occasionally the arguments become somewhat technical, but even in those cases I think that the general thrust of the idea is fairly easy to appreciate. I have sometimes placed short remarks from a more advanced standpoint in square brackets. To help those without a background in molecular biology, I have also included as a frontispiece a figure showing the approximate sizes of molecules, chromosomes, cells, and so forth, as well as two appendixes, the first sketching in the briefest outline the elements of molecular biology and the second setting out the details of the genetic code. Since most people (except chemists) hate chemical formulas, I have banished almost all of them to the first appendix.

In spite of all my efforts at clarification, a layperson may still find parts of chapters 4, 5, and 12 somewhat hard going at first reading. My advice to the reader, should he or she become stuck in such a passage, is either to persevere or to skip to the next chapter. Most of the book is fairly easy. Don't give up hope just because a few paragraphs seem a little hard to follow.

The most important theme of the book is natural selection. As I explain, it is this basic mechanism that makes biology different from all the other sciences. Of course anyone can grasp the mechanism itself, though remarkably few people actually do so. Most surprising, however, are the *results* of such a process, acting over billions of generations. It is the general character of the resulting organisms that is unexpected. Natural selection almost always builds on what went before, so that a basically simple process becomes encumbered with many subsidiary gadgets. As François Jacob has so aptly put it, "Evolution is a tinkerer." It is the resulting complexity that makes biological organisms so hard to unscramble. Biology is thus very different from physics. The basic laws of physics can usually be expressed in exact mathematical form, and they are probably the same throughout the universe. The "laws" of biology, by contrast, are often only broad generalizations, since they describe rather elaborate chemical mechanisms that natural selection has evolved over billions of years.

Biological replication, so central to the process of natural selection, produces many exact copies of an almost infinite variety of intricate chemical molecules. There is nothing like this in physics or its related disciplines. That is one reason why, to some people, biological organisms appear infinitely improbable.

All this can make it difficult for a physicist to contribute to biological research. Elegance and a deep simplicity, often expressed in a very abstract mathematical form, are useful guides in physics, but in biology such intellectual tools can be very misleading. For this reason a theorist in biology has to receive much more guidance from the experimental evidence (however cloudy and confused) than is usually necessary in physics. These arguments are set out in more detail in chapter 13, "Conclusions."

I myself knew very little biology, except in a rather general way, till I was over thirty, since my first degree was in physics. It took me a little time to adjust to the rather different way of thinking necessary in biology. It was almost as if one had to be born again. Yet such a transition is not as difficult as all that and is certainly well worth the effort. To discuss how my career developed, I turn first to a brief account of my early years.

1

Prologue: My Early Years

I WAS BORN IN 1916, in the middle of the First World War. My parents, Harry Crick and Anne Elizabeth Crick (née Wilkins), were a middle-class couple living near the town of Northampton, in the English Midlands. The main industry in Northampton in those days revolved around leather and the manufacture of footwear—so much so that the local soccer team was called the Cobblers. My father, with his eldest brother, Walter, ran a factory, founded by their father, that produced boots and shoes.

I was born at home. I know this because of a curious incident connected with my birth. While my mother was not deeply superstitious, she did like to cultivate certain mildly superstitious practices. Each new year she would try to arrange that the first person who crossed our threshold was dark rather than blond. This practice—I have no idea if it still goes on—is called "first footing" and is supposed to bring good luck in the ensuing year. After I was born she instructed her younger sister, Ethel, to carry me to the top of our house. My mother hoped that this little ceremony would make sure that, in later life, I would "rise to the top." Most superstitious practices reveal more about their perpetrators than they realize, and this family legend shows rather clearly that my mother, like many another mother, was ambitious for her first-born son even before she could have had any inkling of my character and abilities.

I have little recollection of my very early years. I do not even remember being taught to read by my aunt Ethel, who was a schoolteacher. Photographs make me appear to be a very normal child. My mother was fond of saying that I looked like an archbishop. I'm not sure she had ever seen an archbishop—she was not a Catholic or a member of the Church of England—but she may well have seen a photograph of one in the newspaper. It is hardly likely that at the age of four or five I resembled such a venerable person. What I suspect she meant, but was too restrained to say, was that she thought I looked like an angel—very fair hair, blue eyes, an "angelic" expression of benevolent curiosity—but with perhaps something extra. Odile (my current wife) has a locket, a gift from my mother, from that period. It contains two small round tinted photographs, one of my younger brother, Tony, and one of me. I once commented to her that, from the look of it, I seemed to have been a rather angelic child. "Not really," she said. "Look at those piercing eyes." And she spoke with feeling, having often, in our many years together, been subjected to that same critical inquiring gaze.

My only other clue as to my early nature comes from Michael, my son by my first wife, Doreen. When he was about the same age, he lived for a time with my mother. I noticed that, more than once, in reply to an explanation by her, he would answer, "But that can't be right." My mother, puzzled, would ask, "Why not?" to which Michael would give a simple, logical explanation that was transparently correct. I suspect that I too made such remarks to my mother—which was not difficult because she was not a precise thinker—and she found these both disconcerting and fascinating. In any event, it is clear to me now that my mother thought (as many mothers do) that her elder son had exceptional talents, and coming from a solid, middle-class background, she did everything possible to see that these talents were nurtured.

It must have been to parry my constant questions about the world—for neither of my parents had any scientific background—that they bought for me Arthur Mee's *Children's Encyclopedia.* This was published serially, so that in any one number art, science, history, mythology, and literature were all jumbled together. As far as I can remember, I read it all avidly, but it was the science that appealed to me most. What was the universe like? What were atoms? How did things grow? I absorbed great chunks of explanation, reveling in the unexpectedness of it all, judged by the everyday world I saw around me. How marvelous to have discovered such things! It must have been at such an early age that I decided I would be a scientist. But I foresaw one snag. By the time I grew up—and how far away that seemed!—everything would have been discovered. I confided my fears to my mother, who reassured me. "Don't worry, Ducky," she said. "There will be plenty left for you to find out."

By the time I was ten or twelve, I had graduated to experiments at home—my parents must have bought me a student's textbook on chemistry. I tried to make artificial silk—a failure. I put an explosive mixture into bottles and blew them up electrically—a spectacular success that, not unnaturally, worried my parents. A compromise was reached. A bottle could be blown up only while it was immersed in a pail of water. I got a prize at school—my first prize ever—for collecting wildflowers. I had gathered far more species than anyone else, but then we lived on the edge of the country whereas all my fellow schoolboys lived in the town. I felt a little guilty about this but accepted the prize—a small book on insect-eating plants—without demur. I wrote and mimeographed a small magazine to entertain my parents and my friends. But in spite of all this, I do not recall being exceptionally precocious or doing anything really outstanding. I was fairly good at mathematics, but I never discovered for myself some important theorem. In short, I was curious about the world, logical, enterprising, and willing to work hard if my enthusiasm was aroused. If I had a fault, it was that if I could grasp something easily, I believed I had already understood it thoroughly.

My family were all tennis players. My father played for many years for Northamptonshire, an English county, and once played at Wimbledon. My mother also played, but with much less skill and only moderate enthusiasm. My younger brother, Tony, was a much keener player, doing well in the Junior County Championship and also playing for his school. I can hardly believe it now, but as a boy I was mad about tennis. I can still remember the day when my mother woke me early and told me (what bliss!) that I could miss school that day as we were going to Wimbledon. My brother and I would sit, sometimes for hours, beside the courts at the local tennis club, waiting for the drizzle to stop and hoping at least one of the courts would become dry enough for us to play on it. I did play other games (soccer, rugby, cricket, etc.) but without any distinction.

My parents were religious in a rather quiet way. We had nothing like family prayers, but they attended church every Sunday morning and when we were old enough my brother and I went with them. The church was a nonconformist Protestant one, a Congregational Church, as it is called in England, with a substantial building on Abington Avenue. As we did not own an automobile we often walked to church, though sometimes we made part of the journey by bus. My mother greatly admired the clergyman because of his upright character. For a time my father was secretary of the church (that is, he did the church's financial paperwork), but I did not get the feeling that either of them was especially devout. Certainly they were not overly narrow in their outlook on life. My father sometimes played tennis on Sunday afternoons, but my mother warned me not to mention this to other members of the congregation since some almost certainly would not have approved of such sinful conduct.

I accepted all this, as children do, as part of our way of life. At exactly which point I lost my early religious faith I am not clear, but I suspect I was then about twelve years old. It was almost certainly before the actual onset of puberty. Nor can I recall exactly what led me to this radical change of viewpoint. I remember telling my mother that I no longer wished to go to church, and she was visibly upset by this. I imagine that my growing interest in science and the rather lowly intellectual level of the preacher and his congregation motivated me, though I doubt if it would have made much difference if I had known of other more sophisticated Christian beliefs. Whatever the reason, from then on I was a skeptic, an agnostic with a strong inclination toward atheism.

This did not save me from attending Christian services at school, especially at the boarding school I went to later, where a compulsory service was held every morning and two on Sundays. For the first year there, until my voice broke, I even sang in the choir. I would listen to the sermons but with detachment and even with some amusement if they were not too boring. Fortunately, as they were addressed to schoolboys, they were often short, though all too frequently based on moral exhortation.

I have no doubt, as will emerge later, that this loss of faith in Christian religion and my growing attachment to science have played a dominant part in my scientific career, not so much on a day-to-day basis but in the choice of what I have considered interesting and important. I realized early on that it is detailed scientific knowledge which makes certain religious beliefs untenable. A knowledge of the true age of the earth and of the fossil record makes it impossible for any balanced intellect to believe in the literal truth of every part of the Bible in the way that fundamentalists do. And if some of the Bible is manifestly wrong, why should any of the rest of it be accepted automatically? A belief, at the time it was formulated, may not only have appealed to the imagination but also fit well with all that was then known. It can nevertheless be made to appear ridiculous because of facts uncovered later by science. What could be more foolish than to base one's entire view of life on ideas that, however plausible at the time, now appear to be quite erroneous? And what would be more important than to find our true place in the universe by removing one by one these unfortunate vestiges of earlier beliefs? Yet it is clear that some mysteries have still to be explained scientifically. While these remain unexplained, they can serve as an easy refuge for religious superstition. It seemed to me of the first importance to identify these unexplained areas of knowledge and to work toward their scientific understanding, whether such explanations would turn out to confirm existing religious beliefs or to refute them.

Although I found many religious beliefs absurd (the story of the animals in Noah's ark is a good example), I often excused them to myself on the assumption that they originally had some rational basis. This sometimes led me to quite unwarranted assumptions. I was familiar with the account in Genesis in which God makes Eve from one of Adam's ribs. How could such a belief arise? Of course I knew that, at least in certain respects, men were anatomically different from women. What more natural for me to assume that men had one less rib than women? A primitive people, knowing this, could easily believe that this missing rib was used to construct Eve. It never entered my head to check whether this tacit hypothesis of mine corresponded to the facts. It was only some years later, probably when I was an undergraduate, that I let slip to a friend of mine, a medical student, that I understood that women had one more rib than men. To my surprise, instead of agreeing he reacted strongly to this idea and asked me why I thought so. When I explained my reasons he almost fell off his chair with laughter. I learned the hard way that in dealing with myths one should not try to be too rational.

My formal education had few special features. For a number of years I attended the Northampton Grammar School. At the age of fourteen I obtained a scholarship to Mill Hill School in North London, a boys' "public" school (in the English sense, meaning private), consisting mainly of boarders. My father and his three brothers had been at school there. Fortunately the school was good at teaching science and I obtained a thorough grounding in physics, chemistry, and mathematics.

I had a rather vulgar attitude toward pure mathematics, being mainly interested in mathematical results. The exact discipline of rigorous proof held no attraction for me, though I enjoyed the elegance of *simple* proofs. Nor could I feel much enthusiasm for chemistry, which, as then taught to schoolboys, was more like a set of recipes than a science. Much later, when I read Linus Pauling's *General Chemistry,* I found it enthralling. Even so I have never tried to master inorganic chemistry, and my knowledge of organic chemistry is still very patchy. I did enjoy the physics I was taught at school. There was a course in medical biology (the school had a Medical Sixth Form, which prepared pupils for the first Bachelor of Medicine exam), but it never occurred to me to learn about the standard animals of the course: the earthworm, the frog, and the rabbit. I think I must have picked up the elements of Mendelian genetics but I don't think I was ever taught it at school.

I played, or was compelled to play, numerous sports but was rather feeble at all of them except tennis. I managed to be on the school tennis team for my last two years there. When I left school I found I could no longer play it for amusement, so I gave it up and have hardly played it since.

At the age of eighteen I went to University College, London. By that time my parents had moved from Northampton to Mill Hill, so that my younger brother could attend the school as a day boy. I lived at home, going to University College by bus and underground, the journey one way taking the better part of an hour. When I was twenty-one I obtained a second-class Honours degree in physics, with subsidiary mathematics. The teaching in physics had been competent but a shade old-fashioned. We were taught the

Bohr theory of the atom, by then (the mid-1930s) quite out of date. Quantum mechanics was hardly mentioned until a very short course of six lectures at the end of the final year. In the same way, the mathematics I learned was what a previous generation of physicists had found useful. I was taught nothing of eigenvalues or group theory, for example.

Physics has in any case changed beyond recognition since then. At that time there was not even a hint of quantum electrodynamics, let alone quarks or superstrings. Thus, although I was trained in what would now be regarded as historical physics, my current knowledge of modern physics is only at the *Scientific American* level.

After the war I did teach myself the elements of quantum mechanics, but I have never had occasion to use it. Books on this subject were in those days often entitled *Wave Mechanics.* At that time they could be found at the Cambridge University library classified under "Hydrodynamics." No doubt things are different now.

Having obtained my Bachelor of Science degree, I started research at University College, under Professor Edward Neville da Costa Andrade, helped financially by my uncle, Arthur Crick. Andrade put me onto the dullest problem imaginable, the determination of the viscosity of water, under pressure, between 100° and 150° C. I lived in a rented apartment near the British Museum that I shared with an ex-school friend, Raoul Colinvaux, who was studying law.

My main task was to construct a sealable, spherical copper vessel (to hold the water), with a neck that would allow for the expansion of the water. It had to be kept at a constant temperature and its decaying oscillations captured on film. I am no good at precise mechanical construction but I had the help of Leonard Walden, Andrade's senior lab assistant, and an excellent staff in the laboratory workshop. I actually enjoyed making the apparatus, boring though it was scientifically, because it was a relief to be doing something after years of merely learning.

These experiences may have helped me during the war, when I had to devise weapons, but otherwise they were a complete waste of time. What I *had* acquired, however indirectly, was the hubris of the physicist, the feeling that physics as a discipline was highly successful, so why should not other sciences do likewise? I believe this did help me when, after the war, I eventually switched to biophysics. It was a healthy corrective to the rather plodding, somewhat cautious attitude I often encountered when I began to mix with biologists.

When the Second World War started in September 1939, the department was evacuated to Wales. I stayed at home, occupying my time by learning to play squash. My brother (who was then a medical student) taught me on the squash courts at Mill Hill School. The students had been evacuated to Wales, while the school buildings had become an emergency hospital. Tony and I played on a sliding handicap. Every time I lost a game I started the next game with an extra point. If I won a game my advantage was reduced by one point. By the end of the year we were about equal. I played squash occasionally, on and off, for many years, both in London and Cambridge. I always enjoyed it because I never tried to play it seriously. As it is no longer a sensible game for one of my age, I now take my exercise by walking or by swimming in a heated swimming pool in the Southern Californian sunshine.

Eventually, early in 1940, I was given a civilian job at the Admiralty. This enabled me to marry my first wife, Doreen Dodd. Our son Michael was born in London, during an air raid, on November 25, 1940. I worked first in the Admiralty Research Laboratory, next to the National Physical Laboratory in Teddington, a South London suburb. Then I was transferred to the Mine Design Department near Havant, not far from Portsmouth on the south coast of England. After the war ended I was given a job in scientific intelligence at the Admiralty in London. By good fortune a land mine had blown up the apparatus I had so laboriously constructed at University College, so after the war I was not obliged to go back to measuring the viscosity of water.

2

The Gossip Test

DURING MOST of the war I had worked on the design of magnetic and acoustic mines—the noncontact mines—initially under the direction of a well-known theoretical physicist, H. S. W. Massey. Such mines were dropped by our aircraft into shipping channels in the relatively shallow water of the Baltic and the North Sea. There they sat, silently and secretly, on the seabed until they were exploded by an enemy sweep or they blew up one of the enemy ships. The trick in designing their circuits was to make them distinguish in some way between the magnetic fields and sounds of a sweep and those of a ship. In this I had been relatively successful. These special mines were about five times as effective as the standard noncontact mines. After the war it was estimated that mines sank or seriously damaged as many as a thousand enemy merchant vessels.

When the war finally came to an end I was at a loss as to what to do. By that time I was working at the Admiralty Headquarters in Whitehall, in the windowless extension known as The Citadel. I did the obvious thing and applied to become a permanent scientific civil servant. At first they were not sure they wanted me, but eventually, after pressure from the Admiralty and a second interview—the committee was chaired by novelist C. P. Snow—I was offered a permanent job. By this time I was reasonably sure that I didn't want to spend the rest of my life designing weapons, but what did I want to do? I took stock of my qualifications. A not-very-good degree, redeemed somewhat by my achievements at the Admiralty. A knowledge of certain restricted parts of magnetism and hydrodynamics, neither of them subjects for which I felt the least bit of enthusiasm. No published papers at all. The few short Admiralty reports I had written at Teddington would count for very little. Only gradually did I realize that this lack of qualification could be an advantage. By the time most scientists have reached age thirty they are trapped by their own expertise. They have invested so much effort in one particular field that it is often extremely difficult, at that time in their careers, to make a radical change. I, on the other hand, knew nothing, except for a basic training in somewhat old-fashioned physics and mathematics and an ability to turn my hand to new things. I was sure in my mind that I wanted to do fundamental research rather than going into applied research, even though my Admiralty experience would have fit me for developmental work. But did I have the necessary ability?

There was some doubt about this among my friends. Some thought I might do better at scientific journalism—perhaps, one of them suggested, I should attempt to join the staff of *Nature,* the leading scientific weekly. (I don't know what the current editor, John Maddox, would think of this idea.) I consulted mathematician Edward Collingwood, under whom I had worked during the war. As always he was reassuring and helpful. He saw no reason why I should not succeed in pure research. I also asked my close friend Georg Kreisel, now a distinguished mathematical logician. I had run across him when he came, at the age of nineteen, to work in the Admiralty under Collingwood. Kreisel's first paper—an essay on an approach to the problem of mining the Baltic, using the methods of Wittgenstein—Collingwood had wisely locked away in his safe. By this time I knew Kreisel well, so I felt his advice would be solidly based. He thought for a moment and delivered his judgment: "I've known a lot of people more stupid than you who've made a success of it."

Thus encouraged, my next problem was to decide what subject to choose. Since I essentially knew nothing, I had an almost completely free choice. This, as the sixties generation discovered later, only makes the decision more difficult. I brooded over this problem for several months. It was so late in my career that I knew I had to make the right choice the first time. I could hardly try one subject for two or three years and then switch to a radically different one. Whatever choice I made would be final, at least for many years.

Working in the Admiralty, I had several friends among the naval officers. They were interested in science but knew even less about it than I did. One day I noticed that I was telling them, with some enthusiasm, about recent advances in antibiotics—penicillin and such. Only that evening did it occur to me that I myself really knew almost nothing about these topics, apart from what I had read in *Penguin Science* or some similar periodical. It came to me that I was not really telling them about science. I was *gossiping* about it.

This insight was a revelation to me. I had discovered the gossip test—what you are really interested in is what you gossip about. Without hesitation, I applied it to my recent conversations. Quickly I narrowed down my interests to two main areas: the borderline between the living and the nonliving, and the workings of the brain. Further introspection showed me that what these two subjects had in common was that they touched on problems which, in many circles, seemed beyond the power of science to explain. Obviously a disbelief in religious dogma was a very deep part of my nature. I had always appreciated that the scientific way of life, like the religious one, needed a high degree of dedication and that one could not be dedicated to anything unless one believed in it passionately.

By now I was delighted by my progress. I seemed to have found the pass through the interminable mountains of knowledge and could glimpse where I wanted to go. But I still had to decide which of the two areas—we would now call them molecular biology and neurobiology—I should choose. This proved to be much easier. I had little difficulty in convincing myself that my existing scientific background would be more easily applied to the first problem—the borderline between the living and the nonliving—and I decided without further hesitation that that would be my choice.

It should not be imagined that I knew nothing at all of either of my subjects. After the war I had spent a lot of my spare time in background reading. The Admiralty had generously allowed me to go once or twice a week to seminars and courses in theoretical physics at University College during my working hours. Sometimes I would sit at my desk at the Admiralty and surreptitiously read a textbook on organic chemistry. I remembered from my school days a little about hydrocarbons, and even about alcohols and ketones, but what were amino acids? In *Chemical and Engineering News* I read an article by an authority who prophesied that the hydrogen bond would be very important for biology—but what was it? The author had an unusual name—Linus Pauling—but he was quite unknown to me. I read Lord Adrian's little book on the brain and found it fascinating. Also Erwin Schroedinger's *What Is Life?* It was only later that I came to see its limitations—like many physicists, he knew nothing of chemistry—but he certainly made it seem as if great things were just around the corner. I read Hinshelwood's *The Bacterial Cell* but could make little of it. (Sir Cyril Hinshelwood was a distinguished physical chemist, later President of the Royal Society and a Nobel Prize winner.)

In spite of all this reading, I must emphasize that I had only a very superficial knowledge of my two chosen subjects. I certainly had no deep insight into either of them. What attracted me to them was that each contained a major mystery—the mystery of life and the mystery of consciousness. I wanted to know more exactly what, in scientific terms, those mysteries were. I felt it would be splendid if I finally made some small contribution to their solution, but that seemed too far away to worry about.

At this point a crisis suddenly arose. I was offered a job! Not a mere studentship, but an actual job. Hamilton Hartridge, a distinguished but somewhat maverick physiologist, had persuaded the Medical Research Council to set up a small unit for him, to work on the eye. He must have heard I was looking for an opening because he asked me to come to see him. I hastily read his wartime paper on color vision—as I recall, he believed, from his work on the psychology of vision, that there were probably seven types of cones in the eye, not the traditional three. The interview went well and he offered me the job. My problem was that only the week before I had decided that my new field of work was to be molecular biology, not neurobiology.

The decision was a hard one. Finally I told myself that my preference for the living-nonliving borderline had been soundly based, that I would have only one chance to embark on a new career, and that I should not be deflected by the accident of someone offering me a job. Somewhat reluctantly, I wrote to Hartridge and told him that, attractive though it was, I must refuse his offer. Perhaps it was just as well because though I found him a lively and engaging character, he seemed to me a little too bouncy and I was not completely sure we would get on. I also doubt if he would have been very understanding if my work had shown his ideas wrong, as time has proved them to be.

My next task was to find some way of entering my new subject. I went around to University College to see Massey, under whom I had worked during the war, to explain my position and to ask for his help. His first guess when I told him I intended to leave the Admiralty was that I wanted him to get me a job in atomic energy (as it was then called), on which he had worked in Berkeley during the latter stages of the war. He looked surprised when I told him of my interest in biology, but he was very helpful and gave me two valuable introductions. The first was to A. V. Hill, also at University College, a Cambridge physiologist who had made for himself a solid reputation studying the biophysics of muscle, especially the thermal aspects of muscular contraction. For this he had been awarded a Nobel Prize in 1922. He liked the idea that I also should become a biophysicist and perhaps, eventually, work on muscle. He arranged an introduction to Sir Edward Mellanby, the powerful secretary of the Medical Research Council (the MRC). He also gave me some advice. "You should go to Cambridge," he said. "You'll find your own level there."

The second person Massey told me to go to see was Maurice Wilkins. Massey smiled to himself as he said this, and I sensed that Maurice was in some way unusual. They had worked together on isotope separation at Berkeley for the atomic bomb. Wilkins had taken a job under his old boss, John Randall in the physics department at King's College, London, and I went there to see him in the basement rooms in which they all worked.

Randall had persuaded the MRC that they should support the entry of physicists into biology. During the war scientists had acquired much more influence than they had had before it. It was not difficult for Randall, one of the inventors of the magnetron (the crucial development in military applications of radar), to argue that just as physicists had had a decisive influence on the war effort, so they could now turn their hands to some of the fundamental biological problems that lay at the foundations of medical research. Thus there was money available for "biophysics," and the MRC had set up one of its research units at King's College, with Randall as its director.

Exactly what biophysics was, or could usefully become, was less clear. At King's they seemed to feel that an important step would be to apply modern physical techniques to biological problems. Wilkins had been working on a new ultraviolet microscope, using mirrors rather than lenses. Lenses would have had to have been made of quartz, since ordinary glass absorbs ultraviolet light. Exactly what they hoped to discover with these new instruments was less clear, but the feeling was that any new observations made would inevitably lead to new discoveries.

Most of their work involved looking at cells rather than molecules. At this time the full power of the electron microscope had yet to be developed, so observing cells meant accepting the relatively low resolving power of the light microscope. The distance between atoms is more than a thousand times smaller than the wavelength of visible light. Most viruses are far too tiny to be seen in an ordinary high-powered microscope, except perhaps as a minute spot of light against a dark background.

In spite of Maurice's enthusiasm and his very friendly explanations, I was not entirely convinced that this was the right way to go. However, at this stage I knew so little of my new subject that I could form only very tentative opinions. I was mainly interested in the borderline between the living and the nonliving, wherever that was, and most of the work at King's seemed rather far on the biological side of that border.

Perhaps the most useful result of this initial contact was my continued friendship with Maurice. We both had somewhat similar scientific backgrounds. We even looked somewhat alike. Many years later, upon seeing a photograph of Maurice in a textbook that was somewhat confusingly labeled (it was next to one of Jim Watson), a young woman in New York mistook it for one of me, though I was standing in front of her at the time. At one stage I even wondered if we might be distantly related, since my mother's maiden name was Wilkins, but if we are cousins we must be very distant ones. More to the point, we were both of a similar age and traveling the same scientific path from physics to biology.

Maurice did not seem especially unusual to me. Even if I had known, say, that he had a taste for Tibetan music, I doubt if I would have considered that odd. Odile (who became my second wife) thought he was rather strange because when he first arrived for dinner at her apartment in Earl's Court he went straight into the kitchen and lifted the lids of the saucepans to see what was cooking.

She had become accustomed to dealing with naval officers, and they never did things like that. After she discovered that this was not the impertinent curiosity of a hungry man—scientists seemed to be curious about such odd things—but simply that Maurice was interested in cooking, she looked at him in a new light.

My next problem was to decide what to work on and, at least as important, where to do it. I first explored the possibility of working at Birkbeck College in London with the X-ray crystallographer J. D. Bernal. Bernal was a fascinating character. One can get a vivid idea of him by reading C. P. Snow's early science novel *The Search,* since the character Constantine is obviously based on Bernal. It is amusing to note that, in the novel, Constantine wins fame and an F.R.S. by discovering how to synthesize proteins, though Snow wisely didn't indicate exactly what the process was. The plot of the novel turns on the setting up of a biophysics institute, while the final incident concerns the narrator deciding not to expose a fellow scientist for falsifying results and instead to give up his own career in science and become a writer, an incident I suspect modeled on something similar in Snow's career.

When I visited Bernal's laboratory I was discouraged by his secretary, Miss Rimmel, an amiable dragon. "Do you realize," she said, "that people from all over the world want to come to work with the professor? Why do you think he would take you on?" But a more serious difficulty was Mellanby, who said the MRC could not support me if I worked with Bernal. They wanted to see me doing something more biological. I decided to take A. V. Hill's advice and try my luck at Cambridge, if someone there would have me.

I visited the physiologist Richard Keynes, who talked to me as he ate his sandwich lunch in front of his experiment. He was working on ion movement in the giant axon of the squid. I talked to the biochemist Roy Markham, who showed me an interesting result he had recently obtained with a plant virus. Typically he described it in such a cryptic manner (I was not yet familiar with the way nucleic acid absorbed ultraviolet light) that I could not at first grasp what he was telling me. Both were helpful and friendly but neither had any space to offer me. Finally I visited the Strangeways Laboratory, headed by Honor Fell, where they did tissue culture. She introduced me to Arthur Hughes. They had had a physicist at the Strangeways—D. E. Lea—but he had died recently and his room was still vacant. Would I like to work there? The MRC agreed and gave me a studentship. My family also helped me financially so that I had enough to live in lodgings and still had some money to buy books.

I stayed at the Strangeways for the better part of two years. While I was there I worked on a problem they were interested in. Hughes had discovered that chick fibroblasts in tissue culture could engulf, or phagocytose, small crumbs of magnetic ore. Inside the cell these tiny particles could be moved by an applied magnetic field. He suggested I use their movements to deduce something about the physical properties of the cytoplasm, the inside of the cell. I was not deeply interested in this problem but I realized that in a superficial way it was ideal for me, since the only scientific subjects I was fairly familiar with were magnetism and hydrodynamics. In due course this led to a pair of papers, one experimental and one theoretical, in *Experimental Cell Research*—my first published papers. But the main advantage was that the work was not too demanding and left me plenty of time for extensive reading in my new subject. It was then that I began in a very tentative way to form my ideas.

Some time during this period I was asked to give a short talk to some research workers who had come to the Strangeways for a course. I recall the occasion vividly, since I tried to describe to them what the important problems in molecular biology were. They waited expectantly, with pens and pencils poised, but as I continued they put them down. Clearly, they thought, this was not serious stuff, just useless speculation. At only one point did they make any notes, and that was when I told them something factual—that irradiation with X rays dramatically reduced the viscosity of a solution of DNA. I would dearly love to know exactly what I said on that occasion. I *think* I know what I would have said, but the memory is so overlaid with the ideas and developments of later years that I feel I can hardly trust it. Nor, as far as I know, have my notes for the talk survived. However, what I probably discussed was the importance of genes, why one needed to discover their molecular structure, how they might be made of DNA (at least in part), and that the most useful thing a gene could do would be to direct the synthesis of a protein, probably by means of an RNA intermediate.

After a year or so I went to Mellanby to report progress. I told him that I was getting results on the physical properties of cytoplasm but that I had spent much of my time in trying to educate myself. He looked rather skeptical. "What does the pancreas do?" he asked. I had only the vaguest ideas about the function of the pancreas but I managed to mumble something about it producing enzymes, hastily adding that my interests did not lie so much in organs as in molecules. He seemed temporarily satisfied.

I had visited him at a fortunate moment. On his desk lay the papers proposing the establishment of an MRC unit at the Cavendish to study the structure of proteins using the method of X-ray diffraction. It was to be headed by Max Perutz, under the general direction of Sir Lawrence Bragg. To my surprise (because I was still very junior), he asked me what I thought about it. I said I thought it was an excellent idea. I also told Mellanby that now that I had a background in biology, I would like to work on protein structure, since I felt my abilities lay more in that direction. This time he raised no objection, and the way was cleared for me to join Max Perutz and John Kendrew at the Cavendish.

3

The Baffling Problem

It is time to step aside from the details of my career to consider the main problem. Even a cursory look at the world of living things shows its immense variety. Though we find many different animals in zoos, they are only a tiny fraction of the animals of similar size and type. J. B. S. Haldane was once asked what the study of biology could tell one about the Almighty. "I'm really not sure," said Haldane, "except that He must be inordinately fond of beetles." There are thought to be at least 300,000 species of beetles. By contrast, there are only about 10,000 species of birds. We must also take into account all the different types of plants, to say nothing of microorganisms such as yeasts and bacteria. In addition, there are all the extinct species, of which the dinosaurs are the most dramatic example, numbering in all perhaps as many as a thousand times all those alive today.

The second property of almost all living things is their complexity and, in particular, their highly organized complexity. This so impressed our forebears that they considered it inconceivable that such intricate and well-organized mechanisms would have arisen without a designer. Had I been living 150 years ago I feel sure I would have been compelled to agree with this Argument from Design. Its most thorough and eloquent protagonist was the Reverend

William Paley whose book, *Natural Theology—or Evidence of the Existences and Attributes of the Deity Collected from the Appearances of Nature,* was published in 1802. Imagine, he said, that crossing a heath one found on the ground a watch in good working condition. Its design and its behavior could only be explained by invoking a maker. In the same way, he argued, the intricate design of living organisms forces us to recognize that they too must have had a Designer.

This compelling argument was shattered by Charles Darwin, who believed that the *appearance* of design is due to the process of natural selection. This idea was put forward both by Darwin and by Alfred Wallace, essentially independently. Their two papers were read before the Linnean Society on July 1, 1858, but did not immediately produce much reaction. In fact, the president of the society, in his annual review, remarked that the year that had passed had not been marked by any striking discoveries. Darwin wrote up a "short" version of his ideas (he had planned a much longer work) as *The Origin of Species.* When this was published in 1859, it immediately ran through several reprintings and did indeed produce a sensation. As well it might, because it is plain today that it outlined the essential feature of the "Secret of Life." It needed only the discovery of genetics, originally made by Gregor Mendel in the 1860s, and, in this century, of the molecular basis of genetics, for the secret to stand before us in all its naked glory. It is all the more astonishing that today the majority of human beings are not aware of all this. Of those who are aware of it, many feel (with Ronald Reagan) that there must be a catch in it somewhere. A surprising number of highly educated people are indifferent to these discoveries, and in western society a rather vocal minority are actively hostile to evolutionary ideas.

To return to natural selection. Perhaps the first point to grasp is that a complex creature, or even a complex part of a creature, such as the eye, did not arise in one evolutionary step. Rather it evolved through a *series* of small steps. Exactly what is meant by small is not necessarily obvious since the growth of an organism is controlled by an elaborate program, written in its genes. Sometimes a small change in a key part of the program can make a rather large difference. For example, an alteration in one particular gene in *Drosophila* can produce a fruitfly with legs in the place of its antennae.

Each small step is caused by a random alteration in the genetic instructions. Many of these random alterations may do the organism no good (some may even kill it before it is born), but occasionally a particular chance alteration may give that particular organism a selective advantage. This means that in the last analysis the organism will, on average, leave more offspring than it would otherwise do. If this advantage persists in its descendents then this beneficial mutant will gradually, over many generations, spread through the population. In favorable cases every individual will come to possess the improved version of the gene. The older version will have been eliminated. Natural selection is thus a beautiful mechanism for turning rare events (strictly, favorable rare events) into common ones.

We now know—it was first pointed out by R. A. Fisher—that for this mechanism to work inheritance must be "particulate," as first shown by Mendel, and not "blending." In blending inheritance the properties of an offspring are a simple *blend* of those of its parents. In particulate inheritance the genes, which are what is inherited, are particles and do not blend. It turns out that this makes a crucial difference.

For example, in blending inheritance a black animal mated with a white animal would always produce offspring whose color was a blend of black and white, that is, some shade of gray. And *their* offspring, if they bred together, would always remain gray. In particulate inheritance various things can happen. For example, it could be that all the first-generation animals were indeed gray. If these were now mated together, we would obtain in the second generation, *on average,* one-quarter black animals, one-half gray animals, and one-quarter white. [This assumes that color is, in this case, a simple Mendelian character, without dominance.] The genes, being particulate, do not blend, even if their *effects,* in a single animal, blended, so that one white particle (gene) and one black particle, acting together in the same creature, produced a gray animal. This particulate inheritance *preserves* variation (we have mixed black, gray, and white animals after two generations, not just gray ones), whereas blending inheritance *reduces* variation. If inheritance were blending, the offspring of a black animal and a white animal mate, would produce gray animals indefinitely. This is obviously not the case. The fact can be seen clearly in humans: people do not become more and more alike as the generations go on. Variation is preserved.

The Baffling Problem

Darwin, who was a deeply honest man and always faced up to intellectual difficulties, did not know about particulate inheritance and was consequently very disturbed by the criticisms of a Scottish engineer, Fleeming Jenkin. Jenkin pointed out that inheritance (which, without realizing it, Darwin assumed to be blending) would not allow natural selection to work effectively. As particulate inheritance had not yet been thought of, this was a very damning criticism.

What, then, are the basic requirements for natural selection to work? We obviously need something that can carry "information"—that is, the instructions. The most important requirement is that we should have a process for exact replication of this information. It is almost certain that, in any process, some mistakes will be made, but they should occur only rarely, especially if the entity to be replicated carries a lot of information. [In the case of DNA or RNA, the rate of making mistakes, per effective base pair, per generation must, in simple cases, be rather less than the reciprocal of the number of effective base pairs.]

The second requirement is that replication should produce entities that can themselves be copied by the replication process or processes. Replication should not merely be like that of a printing press, when master plates make many copies of a newspaper but each newspaper cannot, by itself, produce further copies of either the press or the newspaper. [In technical terms, replication should be geometrical, not merely arithmetical.]

The third requirement is that mistakes—mutations—should themselves be capable of being copied, so that useful variation can be preserved by natural selection.

There is a final requirement that the instructions and their products should stay together [cross-feeding is to be avoided]. A useful trick is to use a bag—a cell, that is—to do this, but I will not dwell on this point.

In addition, the information needs to do something useful, or to produce other things that will do useful jobs for it, to help it to survive and to produce fertile offspring with a good chance of survival.

In addition to all this, the organism needs sources of raw material (since it has to produce copies of itself), the ability to get rid of waste products, and some sort of source of energy [Free Energy]. All these features are required, but the heart of the matter is obviously the process of exact replication.

This is not the place to explain Mendelian genetics in all its technical details. However, I shall try to provide a glimpse of the astonishing results that a simple mechanism like natural selection can produce over long periods of time. A fuller and very readable account can be found in the early chapters of Richard Dawkins's recent book, *The Blind Watchmaker.* One may wonder at the title of the book. *Watchmaker* obviously refers to the designer that Paley invoked to explain the imaginary watch found on the heath. But why "blind"? I cannot do better than quote Dawkins's actual words:

> All appearances to the contrary, the only watchmaker in nature is the blind forces of physics, albeit deployed in a very special way. A true watchmaker has foresight: he designs his cogs and springs, and plans their interconnections, with a future purpose in his mind's eye. Natural selection, the blind, unconscious, automatic process which Darwin discovered, and which we now know is the explanation for the existence and apparently purposeful form of all life, has no purpose in mind. It has no mind and no mind's eye. It does not plan for the future. It has no vision, no foresight, no sight at all. If it can be said to play the role of watchmaker in nature, it is the *blind* watchmaker (p. 5).

Dawkins gives a very pretty example to refute the idea that natural selection could not produce the complexity we see all around us in nature. The example is a very simple one, but it drives the point home. He considers a short sentence (taken from *Hamlet*):

METHINKS IT IS LIKE A WEASEL.

He first calculates how exceedingly improbable it is that anyone, typing at random (traditionally a monkey, but in his case his eleven-month-old daughter or a suitable computer program) would by chance hit on this exact sentence, with all the letters in their correct place. [The odds turn out to be about 1 in 10^{40}.] He calls this process "single-step selection."

He next tries a different approach, which he calls "cumulative selection." The computer chooses a *random* sequence of twenty-eight letters. It then makes several copies of this but with a certain chance of making random mistakes in the copying. It next proceeds to select the copy that most resembles the target sentence, however slightly. Using this slightly improved version, it then repeats this process of replication (with mutation) followed by selection. In the book Dawkins gives examples of some of the intermediate stages. In one case, after thirty steps, it had produced:

METHINGS IT ISWLIKE B WECSEL

and after forty-three steps it had the sentence completely correct. *How many* steps it takes to do this is partly a matter of chance. In other trials it took sixty-four steps, forty-one steps, and so forth. The point is that by *cumulative* selection one can reach the target in a relatively small number of steps, whereas in *single-step* selection it would take forever.

The example is obviously oversimple, so Dawkins tried a more complex one, in which the computer grew "trees" (organisms) according to certain recursive rules (genes). The results are too complex to reproduce here. Dawkins says: "Nothing in my biologist's intuition, nothing in my 20 years' experience of programming computers, and nothing in my wildest dreams, prepared me for what actually emerged on the screen" (p. 59).

If you doubt the power of natural selection I urge you, to save your soul, to read Dawkins's book. I think you will find it a revelation. Dawkins gives a nice argument to show how far the process of evolution can go in the time available to it. He points out that man, by selection, has produced an enormous variety of types of dog, such as Pekinese, bulldogs, and so on, in the space of only a few thousand years. Here "man" is the important factor in the environment, and it is his peculiar tastes that have produced (by selective breeding, not by "design") the freaks of nature we see preserved all around us as domestic dogs. Yet the time required to do this, on the evolutionary scale of hundreds of millions of years, is extraordinarily short. So we should not be surprised at the ever greater variety of creatures that natural selection has produced on this much larger time scale.

Incidentally, Dawkins's book contains a fair but devastating critique (pages 37–41) of the book *The Probability of God* by Hugh Montefiore, the Bishop of Birmingham. I first knew Hugh when he was Dean of Caius College, Cambridge, and I agree with Dawkins that Hugh's book "... is a sincere and honest attempt, by a reputable and educated writer, to bring natural theology up to date." I also agree wholeheartedly with Dawkins's criticism of it.

At this point I must pause and ask why exactly it is that so many people find natural selection so hard to accept. Part of the difficulty is that the process is very slow, by our everyday standards, and so we rarely have any direct experience of it operating. Perhaps the type of computer game Richard Dawkins describes might help some people to see the power of the mechanism, but not everyone likes to play with computers. Another difficulty is the striking contrast between the highly organized and intricate results of the process—all the living organisms we see around us—and the randomness at the heart of it. But this contrast is misleading since the process itself is far from random, because of the selective pressure of the environment. I suspect that some people also dislike the idea that natural selection has no foresight. The process itself, in effect, does not know where to go. It is the "environment" that provides the direction, and over the long run its effects are largely unpredictable in detail. Yet organisms appear as if they had been designed to perform in an astonishingly efficient way, and the human mind therefore finds it hard to accept that there need be no Designer to achieve this. The statistical aspects of the process and the vast numbers of *possible* organisms, far too many for all but a tiny fraction of them to have existed at all, are hard to grasp. But the process clearly works. All the worries and criticisms just listed have no content when examined carefully, provided the process is understood properly. And we have examples, both from the laboratory and the field, of natural selection in action, from the molecular level to the level of organisms and populations.

I think there are two fair criticisms of natural selection. The first is that we cannot as yet calculate, from first principles, the *rate* of natural selection, except in a very approximate way, though this may become a little easier when we understand in more detail how organisms develop. It is, after all, rather odd that we worry so much how organisms evolved (a process difficult to study, since it happened in the past and is inherently unpredictable), when we still don't know exactly how they work today. Embryology is much easier to study than evolution. The more logical strategy would be to find out first, in considerable detail, how organisms develop and how they work, and only then to worry how they evolved. Yet evolution is so fascinating a subject that we cannot resist the temptation to try to explain it now, even though our knowledge of embryology is still very incomplete.

The second criticism is that we may not yet know all the gadgetry that has been evolved to make natural selection work more efficiently. There may still be surprises for us in the tricks that are used to make for smoother and more rapid evolution. Sex is probably an example of such a mechanism, and there may, for all we know, be others as yet undiscovered. Selfish DNA—the large amounts of DNA in our chromosomes with no obvious function—may turn out to be part of another (see page 147). It is entirely possible that this selfish DNA may play an essential role in the rapid evolution of some of the complex genetic control mechanisms essential for higher organisms.

But leaving these reservations aside, the process is powerful, versatile, and very important. It is astonishing that in our modern culture so few people really understand it.

You could well accept all those arguments about evolution, natural selection, and genes, together with the idea that genes are units of instruction in an elaborate program that both forms the organism from the fertilized egg and helps control much of its later behavior. Yet you might still be puzzled. How, you might ask, can the genes be so clever? What could genes possibly do that would allow the construction of all the very elaborate and beautifully controlled parts of living things?

To answer this we must first grasp what level of size we are talking about. How big is a gene? At the time I started in biology—the late 1940s—there was already some rather indirect evidence suggesting that a single gene was perhaps no bigger than a very large molecule—that is, a macromolecule. Curiously enough, a simple, suggestive argument based on common knowledge also points in this direction.

Genetics tells us that, roughly speaking, we get half of all our genes from our mother, in the egg, and the other half from our father, in the sperm. Now, the head of a human sperm, which contains these genes, is quite small. A single sperm is far too tiny to be seen clearly by the naked eye, though it can be observed fairly easily using a high-powered microscope. Yet in this small space must be housed an almost complete set of instructions for building an entire human being (the egg providing a duplicate set). Working through the figures, the conclusion is inescapable that a gene must be, by everyday standards, very, very small, about the size of a very large chemical molecule. This alone does not tell us what a gene does, but it does hint that it might be sensible to look first at the chemistry of macromolecules.

It was also known at that time that each chemical reaction in the cell was catalyzed by a special type of large molecule. Such molecules were called enzymes. Enzymes are the machine tools of the living cell. They were first discovered in 1897 by Eduard Buchner, who received a Nobel Prize ten years later for his discovery. In the course of his experiments, he crushed yeast cells in a hydraulic press and obtained a rich mixture of yeast juices. He wondered whether such fragments of a living cell could carry out any of its chemical reactions, since at that time most people thought that the cell must be intact for such reactions to occur. Because he wanted to preserve the juice, he adopted a stratagem used in the kitchen: he added a lot of sugar. To his astonishment, the juice fermented the sugar solution! Thus were enzymes discovered. (The word enzyme means "in yeast.") It was soon found that enzymes could be obtained from many other types of cell, including our own, and that each cell contained very many distinct kinds of enzymes. Even a simple bacterial cell may contain more than a thousand different *types* of enzymes. There may be hundreds or thousands of molecules of any one type.

In favorable circumstances an enzyme could be purified away from all the others and its action studied by itself in solution. Such studies showed that each enzyme was very specific, and catalyzed only one particular chemical reaction or, at most, a few related ones. Without that particular enzyme the chemical reaction, under the mild conditions of temperature and acidity usually found in living cells, would proceed only very, very slowly. Add the enzyme and the reaction goes at a good pace. If you make a well-dispersed solution of starch in water, very little will happen. Spit into it and the enzyme amylase in your saliva will start to digest the starch and release sugars.

The next major discovery was that each of the enzymes studied was a macromolecule and that they all belonged to the same family of macromolecules called proteins. The key discovery was made in 1926 by a one-armed American chemist called James Sumner. It is not all that easy to do chemistry when you have only one arm (he had lost the other in a shooting accident when he was a boy) but Sumner, who was a very determined man, decided he would nevertheless demonstrate that enzymes were proteins. Though he showed that one particular enzyme, urease, was a protein and obtained crystals of it, his results were not immediately accepted. In fact, a group of German workers hotly contested the idea, which somewhat embittered Sumner, but it turned out that he was correct. In 1946 he was awarded part of the Nobel Prize in Chemistry for his discovery. Though very recently a few significant exceptions to this rule have turned up, it is still true that almost all enzymes are proteins.

Proteins are thus a family of subtle and versatile molecules. As soon as I learned about them I realized that one of the key problems was to explain how they were synthesized.

There was a third important generalization, though in the 1940s this was sufficiently new that not everybody was inclined to accept it. This idea was due to George Beadle and Ed Tatum. (They too were to receive a Nobel Prize, in 1958, for their discovery.) Working with the little bread-mold *Neurospora,* they had found that each mutant of it they studied appeared to lack just a single enzyme. They coined the famous slogan "One gene—one enzyme."

Thus the general plan of living things seemed almost obvious. Each gene determines a particular protein. Some of these proteins are used to form structures or to carry signals, while many of them are the catalysts that decide what chemical reactions should and should not take place in each cell. Almost every cell in our bodies has a complete set of genes within it, and this chemical program directs how each cell metabolizes, grows, and interacts with its neighbors. Armed with all this (to me) new knowledge, it did not take much to recognize the key questions. What are genes made of? How are they copied exactly? And how do they control, or at least influence, the synthesis of proteins?

It had been known for some time that most of a cell's genes are located on its chromosomes and that chromosomes were probably made of nucleoprotein—that is, of protein and DNA, with perhaps some RNA as well. In the early 1940s it was thought, quite erroneously, that DNA molecules were small and, even more erroneously, simple. Phoebus Levene, the leading expert on nucleic acid in the 1930s, had proposed that they had a regular repeating structure [the so-called tetranucleotide hypothesis]. This hardly suggested that they could easily carry genetic information. Surely, it was thought, if genes had to have such remarkable properties, they must be made of proteins, since proteins as a class were known to be capable of such remarkable functions. Perhaps the DNA there had some associated function, such as acting as a scaffold for the more sophisticated proteins.

It was also known that each protein was a polymer. That is, it consisted of a long chain, known as a polypeptide chain, constructed by stringing together, end to end, small organic molecules, called monomers since they are the elements of a polymer. In a homopolymer, such as nylon, the small monomers are usually all the same. Proteins are not as simple as that. Each protein is a heteropolymer, its chains being strung together from a selection of somewhat different small molecules, in this instance amino acids. The net result is that, chemically speaking, each polypeptide chain has a completely regular backbone, with little side-chains attached at regular intervals. It was believed that there were about twenty different possible side-chains (the exact number was not known at that time). The amino acids (the monomers) are just like the letters in a font of type. The base of each kind of letter from the font is always the same, so that it can fit into the grooves that hold the assembled type, but the top of each letter is different, so that a particular letter will be printed from it. Each protein has a characteristic number of amino acids, usually several hundred of them, so any particular protein could be thought of crudely as a paragraph written in a special language having about twenty (chemical) letters. It was not then known for certain, as it is now, that for each protein the letters have to be in a particular order (as indeed they have to be in a particular paragraph). This was first shown a little later by the biochemist Fred Sanger, but it was easy enough to guess that this was likely to be true.

Of course each paragraph in our language is really one long line of letters. For convenience this is split up into a series of lines, written one under the other, but this is only a secondary matter, since the meaning is exactly the same whether the lines are long or short, few or many, provided we take care about splitting the words at the end of each line. Proteins were known to be very different. Although the polypeptide backbone is chemically regular, it contains flexible links, so that in principle many different three-dimensional shapes are possible. Nevertheless, each protein appeared to have its own shape, and in many cases this shape was known to be fairly compact (the word used was "globular") rather than very extended (or "fibrous"). A number of proteins had been crystallized, and these crystals gave detailed X-ray diffraction patterns, suggesting that the three-dimensional structure of each molecule of a particular kind of protein was exactly (or almost exactly) the same. Moreover many proteins, if heated briefly to the boiling point of water, or even to some temperature below this, became denatured, as if they had unfolded so that their three-dimensional structure had been partly destroyed. When this happened the denatured protein usually lost its catalytic or other function, strongly suggesting that the function of such a protein *depended on its exact three-dimensional structure.*

And now we can approach the baffling problem that appeared to face us. If genes are made of protein, it seemed likely that each gene had to have a special three-dimensional, somewhat compact structure. Now, a vital property of a gene was that it could be copied exactly for generation after generation, with only occasional mistakes. What we were trying to guess was the general nature of this copying mechanism. Surely the way to copy something was to make a complementary structure—a mold—and then to make a further complementary structure of the mold, to produce in this way an exact copy of the original. This, after all, is how, broadly speaking, sculpture is copied. But then the dilemma arose: It is easy to copy the *outside* of a three-dimensional structure in this way, but how on earth could one copy the *inside?* The whole process seemed so utterly mysterious that one hardly knew how to begin thinking about it.

Of course, now that we know the answer, it all seems so completely obvious that no one nowadays remembers just how puzzling the problem seemed then. If by chance you do *not* know the answer, I ask you to pause a moment and reflect on what the answer might be. There is no need, at this stage, to bother about the details of the chemistry. It is the principle of the idea that matters. The problem was not made easier by the fact that many of the properties of proteins and genes just outlined were not known for certain. All of them were plausible and most of them seemed very probable but, as in most problems near the frontiers of research, there were always nagging doubts that one or more of these assumptions might be dangerously misleading. In research the front line is almost always in a fog.

So what was the answer? Curiously enough, I had arrived at the correct solution before Jim Watson and I discovered the double-helical structure of DNA. The basic idea (which was not entirely new) was this: All a gene had to do was to get the *sequence* of the amino acids correct in that protein. Once the correct polypeptide chain had been synthesized, with all its side chains in the right order, then, following the laws of chemistry, the protein *would fold itself up correctly into a unique three-dimensional structure.* (What the exact three-dimensional structure of each protein was remained to be determined.) By this bold assumption the problem was changed from a three-dimensional one to a one-dimensional one, and the original dilemma largely disappeared.

Of course, this had not *solved* the problem. It had merely transformed it from an intractable one to a manageable one. For the problem still remained: how to make an exact copy of a one-dimensional sequence. To approach that we must return to what was known about DNA.

By the late 1940s our knowledge of DNA had improved in several important respects. It had been discovered that DNA molecules were not, after all, very short. Exactly how long they were was not clear. We know now that they appeared to be short because, being long molecules (in the sense that a piece of string is long), they could easily be broken in the process of getting them out of the cell and manipulating them in the test tube. Just stirring a DNA solution is enough to break the longer molecules. Their chemistry was now known more correctly, and moreover the tetranucleotide hypothesis was dead, killed by some very beautiful work by a chemist at Columbia, the Austrian refugee Erwin Chargaff. DNA was known to be a polymer, but with a very different backbone and with only four letters in its alphabet, rather than twenty. Chargaff showed that DNA from different sources had rather different amounts of those four bases (as they were called). Perhaps DNA was not such a dumb molecule after all. It might conceivably be long enough and varied enough to carry some genetic information.

Even before I left the Admiralty there had been some quite unexpected evidence pointing to DNA as near the center of the mystery. In 1944 Avery, MacLeod, and McCarty, who worked at the Rockefeller Institute in New York, had published a paper claiming that the "transforming factor" of pneumococcus consisted of pure DNA. The transforming factor was a chemical extracted from a strain of bacteria having a smooth coat. When added to a related strain lacking such a coat it "transformed" it, so that some of the recipient bacteria acquired the smooth coat. More important, all the descendants of such cells had the same smooth coat. In the paper the authors were rather cautious in interpreting their result, but in a now-famous letter to his brother Avery expressed himself more freely. "Sounds like a virus—may be a gene," he wrote.

This conclusion was not immediately accepted. An influential biochemist, Alfred Mirsky, also at the Rockefeller, was convinced that it was an impurity of the DNA that was causing the transformation. Subsequently more careful work by Rollin Hotchkiss at the Rockefeller showed that this was highly unlikely. It was argued that Avery, MacLeod, and McCarty's evidence was flimsy, in that only one character had been transformed. Hotchkiss showed that another character could also be transformed. The fact that these transformations were often unreliable, tricky to perform, and only altered a minority of cells did not help matters. Another objection was that the process had been shown to occur just in these particular bacteria. Moreover, at that time no bacterium of any sort had been shown to have genes, though this was discovered not long afterward by Joshua Lederberg and Ed Tatum. In short, it was feared that transformation might be a freak case and misleading as far as higher organisms were concerned. This was not a wholly unreasonable point of view. A single isolated bit of evidence, however striking, is always open to doubt. It is the accumulation of several different lines of evidence that is compelling.

It is sometimes claimed that the work of Avery and his colleagues was ignored and neglected. Naturally there was a mixed spectrum of reactions to their results, but one can hardly say no one knew about it. For example, that august and somewhat conservative body, the Royal Society of London, awarded the Copley Medal to Avery in 1945, specifically citing his work on the transforming factor. I would dearly love to know who wrote the citation for them.

Nevertheless, even if all the objections and reservations are brushed aside, the fact that the transforming factor was pure DNA does not in itself prove that DNA alone is the genetic material in pneumococcus. One could quite logically claim that a gene there was made of DNA *and* protein, each carrying part of the genetic information, and it was just an accident of the system that in transformation the altered DNA part was carrying the information to change the polysaccharide coat. Perhaps in another experiment a protein component might be found that would also produce a heritable change in the coat or in other cell properties.

Whatever the interpretation, because of this experiment and because of the increased knowledge of the chemistry of DNA, it was now possible that genes might be made of DNA alone. Meanwhile the main interest of the group at the Cavendish was in the three-dimensional structure of proteins such as hemoglobin and myoglobin.

4

Rocking the Boat

LET US NOW return to my own career. I still had to make contact with Max Perutz. One day in the late 1940s, I was returning to Cambridge from a visit to London, having arranged to call on Perutz at the physics laboratory where he worked. The train journey from London was uneventful. I watched the countryside slide past but my thoughts were elsewhere, focused mainly on my impending visit to the Cavendish Laboratory. For a British physicist the Cavendish had a unique glamour. It had been named after the eighteenth-century physicist Henry Cavendish, a recluse and an experimenter of genius. The first professor had been the Scottish theoretical physicist James Clerk Maxwell, of Maxwell's equations. While the laboratory was being built he did experiments in his kitchen at home, his wife raising the room temperature for him by boiling pans of water.

It was at the Cavendish that J. J. Thomson had "discovered" the electron by making measurements of both its mass and its charge. Thompson was an interesting case of an experimenter who was so clumsy that his associates tried to keep him away from his own apparatus, for fear of his breaking it. Ernest Rutherford, fresh from New Zealand, had started his main research career there and later returned to succeed J.J. as Cavendish Professor. There, under his direction Cockroft and Walton had first "smashed the atom"—that is, had produced the first artificial atomic disintegration. Their original accelerator was still there. And in the early 1930s James

Chadwick (whom I knew later as Master of Caius College) had in a few short weeks discovered the neutron. At that time the Cavendish was in the very forefront of research in fundamental physics.

The current Cavendish Professor was Sir Lawrence Bragg (known to his close friends as Willie), the formulator of Bragg's law for X-ray diffraction. He was the youngest Nobel Prize winner ever, having been only twenty-five when he shared it with his father, Sir William Bragg. It was no wonder that I was in awe of such a world-famous institution and excited at the prospect of visiting it.

At the station I decided to take a taxi. After settling my bags, I leaned back in my seat. "Take me," I said, "to the Cavendish Laboratory."

The driver turned his head to look at me over his shoulder. "Where's that?" he asked.

I realized, not for the first time, that not everyone was as deeply interested in fundamental science as I was. After fumbling in my papers I found the address.

"It's in Free School Lane," I said, "wherever that is."

"Not far from the Market Square," said the cabby, and off we went.

Max Perutz, whom I was to visit, was Austrian by birth. He had obtained his first degree, in chemistry, at the University of Vienna. He had wanted to go to Cambridge to work under Gowland Hopkins, the founder of the Cambridge School of Biochemistry. Perutz had asked Herman Mark, the polymer specialist, to try to arrange this for him when Mark went on a short visit to Cambridge. Instead Mark ran into J. D. Bernal (known to his close friends as "Sage," because he appeared to know everything). Bernal said he would be happy to have Perutz work with him and so Max became a crystallographer. This was all before the Second World War.

By the time of my visit Perutz was working, under the loose supervision of Bragg, on the three-dimensional structures of proteins. As I explained in the last chapter, proteins belong to one of the key families of biological macromolecules. How each protein acts depends on its exact three-dimensional structure. It is therefore crucially important to discover such structures experimentally. At that time the largest organic molecule whose three-dimensional structure had been determined by X-ray diffraction was two orders of magnitude smaller than a typical protein. A determination of the three-dimensional structure of a protein seemed, to most crystallographers, almost impossible or, at best, very far away. Bernal had always been enthusiastic about it, but then he was a visionary. However, it also had a great appeal for the hard-headed Bragg, since it represented a challenge. Having started his career unraveling the very simple structure of crystals of sodium chloride (common table salt), Bragg hoped he might crown his achievements by solving one of the *largest* possible molecular structures.

Before the war, Bernal had founded the study of the X-ray diffraction of protein crystals. One day, he was observing the optical properties of a protein crystal, using the light microscope (actually a polarizing microscope). The crystal was sitting on an open glass slide, with a little bit of the mother liquor of the crystal (the solution in which the protein crystal had been grown) attached to it. Slowly the water in the mother liquor evaporated into the air till eventually the crystal became dry. As it did so Bernal saw the optical properties deteriorate, since the dry jumbled crystal transmitted the light in a more confused way than before. Bernal immediately realized that it was important to keep protein crystals wet and proceeded to mount a crystal in a small silicon tube, sealed with a special wax at each end. Fortunately the silicon interfered very little with the X rays being diffracted from the crystal. All previous attempts to get X-ray diffraction photos from protein crystals had produced only a few smudges on the photographic plate since the crystals used had dried in the air. Great was the excitement in Bernal's lab when the wet crystal produced many beautiful spots. The study of protein structure had taken a decisive first step.

Before I first visited Max Perutz at the Cavendish I read the two papers he had recently published in the *Proceedings of the Royal Society* about his X-ray diffraction studies on crystals of a variety of hemoglobin. Hemoglobin is the protein that carries oxygen in our blood and makes red blood cells red, though the variety Perutz had studied came from a horse, as horse hemoglobin happens to form crystals that are especially convenient for X-ray studies. We now know that each hemoglobin molecule is made up of four rather similar subunits, each of which contains about 2,500 atoms, arranged in a precise three-dimensional structure.

Since one cannot easily focus X rays, it is impossible to make X-ray photographs in the way one uses a lens to make photographs using visible light or by focusing electrons in the electron microscope. However, the wavelength of convenient X rays is about the same distance as the distance between close atoms in an organic molecule. For this reason the pattern of X rays that molecules scatter can, under optimal circumstances, contain enough information for the experimenter to determine the positions of all the atoms in the molecule. More correctly, such a picture shows the density of the electrons that surround each atom and that, since they have very little mass, scatter the X rays more effectively than the heavier atomic nuclei. A crystal is used because the X rays scattered from a single molecule would be too feeble. If long exposures were used to try to overcome this difficulty, the heavy dose of X rays would damage the molecule far too much and effectively cook it before enough X rays had been scattered to be useful.

In those days the X rays were registered by special photographic film, developed in much the same way that ordinary photographic negatives are developed. Nowadays the X rays are caught and measured by counters. A special camera had to move the crystal in the beam, and the X-ray film with it, in order to record a particular portion of the diffraction data at a time.

Although I must have learned all this when I took my B.Sc. in physics, I had forgotten most of it by this time, so that I could get only a rough idea of what Perutz had been doing. I learned that protein crystals usually had a lot of water in them, tucked away in the interstices in the crystal between one large molecule and its neighbors. In a drier atmosphere a crystal could shrink somewhat, as the protein molecules packed more closely together, and it was these shrinkage stages that Perutz had been studying. If the atmosphere were *too* dry, the packing of the molecules would become jumbled, as the bulky molecules vainly tried to get as near together as possible. The nice X-ray diffraction pattern, with many sharp discrete spots, would then deteriorate to a few smudges on the X-ray film. In diffraction, *regular* three-dimensional structures produce a whole series of discrete spots, as Bragg had explained many years before.

I also knew about the major problem of X-ray crystallography. Even if the strength of all the many X-ray spots were measured (in those days a tremendous undertaking) and even if the atoms in the crystals were so regular that those X-ray spots corresponding to fine details were also recorded, the mathematics showed clearly that the spots contained just half the information to reveal the three-dimensional structure. [In technical terms the spots gave the intensities of all the many Fourier components but not their phases.] If by some magic the position of each atom were known, then it was possible (though in those days very laborious) to calculate exactly what the X-ray diffraction pattern would look like, and also to calculate the missing information—the phases. But given only the spots, the theory showed that very, very many possible three-dimensional arrangements of electron density could give exactly the same spots, and there was no easy way to decide which was the correct one.

In recent years it has been shown, mainly by the work of Jerome Karle and Herbert Hauptman, how to do this for small molecules by putting various rather natural constraints into the mathematics. For this work they were awarded the Nobel Prize for Chemistry in 1985. But even today such methods cannot, by themselves, be used for large molecules of the size of most proteins.

Thus it was not surprising that in the late 1940s Perutz had not progressed very far. I listened carefully to his explanation of his work and even ventured a few comments. This must have made me appear more perceptive and quicker on the uptake than I really was. In any event, I impressed Perutz sufficiently for him to welcome the idea of my joining him, provided the MRC would support me.

In 1949 Odile and I got married. We had first met during the war when she was a naval officer—strictly speaking a WREN officer (the British equivalent of the WAVES, the women's naval service). Toward the latter part of the war she worked at the Admiralty headquarters in Whitehall (the main government street in London), translating captured German documents. After the war she became an art student again, this time at St. Martin's School of Art, in Charing Cross Road, not far from Whitehall. I was then working in Whitehall myself, in Naval Intelligence, so it was easy for us to meet. In 1947 Doreen and I were divorced. Odile had transferred to a new course in fashion design at the Royal College of Art, but after the first year she decided she preferred marriage to further study.

We spent our honeymoon in Italy. Only after we returned did I discover that the First International Congress of Biochemistry had taken place in Cambridge while we were away. In those days there were nothing like as many scientific meetings as there are now. As a beginner in research, still almost an amateur, I was not especially aware of even those meetings that did take place. I think at the back of my mind was the idea that science was an occupation for gentlemen (even if somewhat impoverished gentlemen). Incredible as it may seem, I had not realized that for many it was a highly competitive career.

The Perutzes had lived for some time in a tiny furnished apartment very conveniently located near the center of Cambridge and only a few minutes' walk from the Cavendish. They now planned to move into a suburban house, to have more room, and suggested to us that we might take their place. We were delighted with the idea and moved into The Green Door, as it was called, a set of two and one-half rooms and a small kitchen, at the top of The Old Vicarage, next to St. Clement's church on Bridge Street, between the top of Portugal Place and Thompson's Lane. The owner, a tobacconist, and his wife lived in the main body of the house while we occupied the attic. The actual Green Door was on the ground floor, at the back, leading to a narrow staircase that went up to our set of rooms. The washbasin and lavatory were halfway up these stairs and the bath, covered by a hinged board, was tucked into the small kitchen. It was often necessary to move a miscellaneous collection of saucepans and dishes if one wanted to have a bath. One room served as a living room, the other as a bedroom, while the smallest room was used as a bedroom for my son Michael, when he came home for the holidays from his boarding school.

Odile and I had our leisurely breakfasts by the attic window in the little living room, looking out over the graveyard to Bridge Street and beyond that to the chapel of St. John's College. There was much less motor traffic in those days, though many bicycles. Sometimes in the evening we would hear an owl hooting from one of the trees that bordered the college. We had only a small income but fortunately the rent was also very small, even though the apartment was rented furnished. The landlord apologized profusely when he felt compelled to raise our rent from thirty shillings a week to thirty shillings and sixpence. Odile luxuriated in her newly found leisure, read French novels in front of the small gas fire, and attended, informally, a few lectures on French literature, while I reveled in the romance of doing real scientific research and in the fascination of my new subject.

The first thing I had to do was to teach myself X-ray crystallography, both the theory and the practice. Perutz advised me which textbooks to read and I was shown the elements of mounting crystals and taking X-ray pictures. Simple inspection of parts of the X-ray diffraction pattern usually gave, in a fairly straightforward manner, not only the physical dimensions of the unit cell—the spatial repeat unit—but also revealed something about its symmetry. Because biological molecules often have a "handedness"—their mirror image is not usually found in living things—certain symmetry elements [inversion through a center, reflection, and the related glide planes] cannot occur in protein crystals. This limitation reduces drastically the possible number of symmetry combinations, or space groups, as they are called.

There is also a well-known limitation on rotation axes. Wallpaper can have a twofold rotation axis—it looks exactly the same if it is rotated by 180 degrees—or a threefold, fourfold, or sixfold one. All other rotational axes are impossible, including a fivefold one. This restriction is true for any extended pattern with two-dimensional symmetry, known as a plane group, and thus also for three-dimensional extended symmetry, or a space group. Of course a *single* object can have fivefold symmetry. The regular dodecahedron and icosahedron, which have fivefold rotational axes, were known to the Greeks, but what is allowed for a point group (which has no dimensions) is impossible for a plane group (of two dimensions) or a space group (of three dimensions). Moslem art, which for religious reasons is forbidden to depict people or animals (since the Prophet was very hostile to paganism), is often for this reason very geometrical in design. One can sometimes see the artist flirting with local fivefold symmetry without ever attaining it on a repeating basis. As it turns out, the protein shells of many small "spherical" viruses (such as the polio virus) usually have fivefold symmetry, but that is another story.

The theory of the X-ray diffraction of crystals is straightforward, so much so that most modern physicists find it rather dull. Although it is necessary to be able to handle the algebraic details, I soon found I could see the answer to many of these mathematical problems by a combination of imagery and logic, without first having to slog through the mathematics.

Some years later, when Jim Watson joined us at the Cavendish, I used some of these visual methods, based on the deeper mathematics, to teach him the outlines of X-ray diffraction. I even considered writing a small didactic monograph on it, to be entitled

"Fourier Transforms for Bird Watchers" (Jim had become a biologist because of an early interest in bird-watching), but there were too many other distractions and I never wrote it.

At that time there was no easily available textbook along these lines. The existing texts usually used a step-by-step method, based largely on Bragg's law and the historical development of the subject. To someone like myself this only made it more difficult and certainly more tedious, since an elementary method often arouses deeper questions in the learner and these worries can impede one's progress in learning. It is often better, at least for the brighter pupils, to go straight to the advanced treatment and try to get over the more powerful formalism while at the same time attempting to provide some insight into what is going on. In my case there was no alternative but to teach X-ray diffraction to myself. This was useful as I acquired a fairly thorough and intimate knowledge of it. Moreover, because Perutz was studying the shrinkage stages of a crystal made of large molecules, I learned how to deal with diffraction from a *single* molecule, and only then arranged them in a regular crystal lattice, rather than following the more conventional path of starting off with them in a lattice. This proved valuable to me later.

Armed with this new knowledge, I reread Perutz's papers and spent some time thinking about how the problem of protein structure was to be solved. Perutz had tentatively suggested that the shape of the molecule was somewhat like an old-fashioned lady's hatbox, and he had put such a diagram into his first paper. (Incidentally, diagrams of models are often difficult to draw satisfactorily, since, unless care is taken, they usually convey more than one intends.) For various reasons I thought that the hatbox was implausible, and I tried to find evidence for other possible shapes. Remember that the relevant X-ray data could not by itself tell us the shape, but that any proposed shape could be used to calculate the X-ray data. The shape influences only the few X-ray reflections that correspond to the coarse structure of the crystal. Their strength depends on the contrast between the high electron density of the protein and the lower electron density of the "water" (actually a salt solution) in between the molecules. Even if such a low-resolution picture of the electron density were available, it would not immediately give the shape of a single molecule, since at various places the protein molecules are in close contact. Where one molecule finished and the next began could not be seen. Fortunately Perutz had studied a set of similar packings—the several shrinkage stages—and by assuming that protein molecules are relatively rigid and merely packed together a little differently in the different stages, the range of possible shapes could be restricted.

I made some progress with the main problem but eventually became stuck. Meanwhile Bragg had independently thought about it. Whereas I had gotten bogged down, he made rapid progress. He boldly assumed that one could approximate the shape by an ellipsoid—a particularly simple type of distorted sphere. Then he looked at what little was known of the crystals of hemoglobin of other species of animal, on the assumption that all types of hemoglobin molecules were likely to have about the same shape. Moreover, he was not disturbed if the data did not *exactly* fit his model, since it was unlikely that the molecule was *exactly* an ellipsoid. In other words he made bold, simplifying assumptions; looked at as wide a range of data as possible; and was critical but not pernickety, as I had been, about the fit between his model and experimental facts. He arrived at a shape that we now know is not a bad approximation to the molecule's real shape, and he and Perutz published a paper on it. The result was not of first-class importance, if only because the method was indirect and needed confirmation by more direct methods, but it was a revelation to me as to how to do scientific research and, more important, how *not* to do it.

As I learned more about the main problem, I began to worry about how it might be solved. As I have said, the X-ray data contained just half the necessary information, though it was known that some of what was available was probably redundant. Was there any systematic way to use the available data? It turned out there was. Some years earlier a crystallographer, Lindo Patterson, had shown that experimental data could be used to construct a special density map, now called a Patterson. [All the amplitudes of the Fourier components are squared and all the phases are put to zero.]

What did this density map mean? Patterson showed that it represented all the possible *interpeak* distances in the real electron density map, all superimposed, so that if the real density map frequently had high density a distance of 10 Å apart in a certain direction, then there would be a peak at 10 Å from the origin in the appropriate direction in the Patterson map. (One Ångstrom unit is equal to one ten-billionth of a meter.) In mathematical terms, this would be a three-dimensional map of the autocorrelation function of the electron density. For a unit cell with very few atoms in it, and using high-resolution X-ray data, one could sometimes unscramble this map of all the possible interatomic distances and obtain the real map of the atomic arrangements. Alas, for protein there were far too many atoms and the resolution was too poor, so that doing this was quite hopeless. Nevertheless, strong features in the Patterson could hint at broad features in the atomic arrangements, and indeed Perutz had predicted that the protein was folded to give rods of electron density, lying in a particular direction, because he saw rods of high density in that direction in the Patterson. As it turned out the latter rods were not really as high as he had imagined (he had at that time only the relative intensity of his X-ray spots, not their absolute value) so the folding was not quite as simple as he had conjectured.

This calculation of the Patterson of his crystals of horse hemoglobin was a difficult and laborious piece of work, since in those days the methods, both for collecting X-ray data and for calculating Fourier Transforms, were, by modern standards, primitive in the extreme. Many crystals had to be mounted (since each would only take a certain dose of X rays before deteriorating); many X-ray photos had to be taken, cross-calibrated, measured by eye, and systematic corrections made. The calculations were not done on what we would now call a computer (that came later) but using an IBM punched card machine. They took an assistant three months and were very laborious. Then all the numbers obtained had to be plotted and contours drawn, till eventually one ended up with a stack of transparent sheets, each having a section of the Patterson density shown on them as contours. As I recall, the negative contours (the average correlation was taken as zero) were omitted and only the positive ones plotted.

I received another lesson when Perutz described his results to a small group of X-ray crystallographers from different parts of Britain assembled in the Cavendish. After his presentation, Bernal rose to comment on it. I regarded Bernal as a genius. For some reason I had acquired the idea that all geniuses behaved badly. I was therefore surprised to hear him praise Perutz in the most genial way for his courage in undertaking such a difficult and, at that time, unprecedented task and for his thoroughness and persistence in carrying it through. Only then did Bernal venture to express, in the nicest possible way, some reservations he had about the Patterson method and this example of it in particular. I learned that if you have something critical to say about a piece of scientific work, it is better to say it firmly but nicely and to preface it with praise of any good aspects of it. I only wish I had always stuck to this useful rule. Unfortunately I have sometimes been carried away by my impatience and expressed myself too briskly and in too devastating a manner.

It was at such a seminar that I gave my first crystallographic talk. Although I was over thirty it was only the second research seminar I had ever given, the first having been about moving magnetic particles in cytoplasm. I made the usual beginner's mistake of trying to get too much into the allotted twenty minutes and was disconcerted to see, after I was about halfway through, that Bernal was fidgeting and only half paying attention. Only later did I learn that he was worrying about where his slides were for the talk he was to give following mine.

All this was of little consequence compared to the subject of my talk, which, broadly speaking, was that they were all wasting their time and that, according to my analysis, almost all the methods they were pursuing had no chance of success. I went through each method in turn, including the Patterson, and tried to demonstrate that all but one was quite hopeless. The exception was the so-called method of isomorphous replacement, which I had calculated had some prospect of success, provided it could be done chemically.

As I mentioned earlier, X-ray diffraction data normally gives us only half the information we need to reconstruct the three-dimensional picture of the electron density of a crystal. We need this three-dimensional picture to help us locate the many thousands of atoms in the crystal. Is there any means of obtaining the missing part of the data? It turns out there is. Suppose a very heavy atom, such as mercury, can be added to the crystal at the same spot on every one of the protein molecules it contains. Suppose this addition does not disturb the packing together of the protein molecules but only displaces an odd water molecule or two. We can then obtain two different X-ray patterns: one without the mercury there, and one with it. By studying the *differences* between the two patterns we can, with luck, locate where the mercury atoms lie in the crystal [strictly, in the unit cell]. Having found these positions, we can obtain some of the missing information by seeing, for each X-ray spot, whether the mercury has made that spot weaker or stronger.

This is the so-called method of isomorphous replacement. "Replacement," because we have replaced a light atom or molecule, such as water, with a heavy atom, such as mercury, which diffracts the X rays more strongly. "Isomorphous," because the two protein crystals—one with the mercury and one without—should have the same form [for the unit cell]. In a loose way, we can think of the added heavy atom as representing a locatable marker to help us find our way among all the other atoms there. It turns out that we usually need at least two *different* isomorphous replacements to allow us to retrieve most of the missing information, and preferably three or more.

This well-known method had already been used successfully to help solve the structure of small molecules. There had previously been one or two halfhearted attempts to use it on proteins, but these had failed, probably because the chemistry used was too crude. Nor was I helped by my title. I had told John Kendrew the sort of thing I intended to say and asked him what I should call it. "Why not," he said, "call it 'What Mad Pursuit'!" (a quotation from Keats' "Ode on a Grecian Urn")—which I did.

Bragg was furious. Here was this newcomer telling experienced X-ray crystallographers, including Bragg himself, who had founded the subject and been in the forefront of it for almost forty years, that what they were doing was most unlikely to lead to any useful result. The fact that I clearly understood the theory of the subject and indeed was apt to be unduly loquacious about it did not help. A little later I was sitting behind Bragg, just before the start of a lecture, and voicing to my neighbor my usual criticism of the subject in a rather derisive manner. Bragg turned around to speak to me over his shoulder. "Crick," he said, "you're rocking the boat."

There was some justification for his annoyance. A group of people engaged in a difficult and somewhat uncertain undertaking are not helped by persistent negative criticism from one of their number. It destroys the mood of confidence necessary to carry through such a hazardous enterprise to a successful conclusion. But equally it is useless to persist in a course of action that is bound to fail, especially if an alternative method exists. As it has turned out, I was completely correct in all my criticisms with one exception. I underestimated the usefulness of studying simple, repeating, artificial peptides (distantly related to proteins), which before long was to give some useful information, but I was quite correct in predicting that only the isomorphous replacement method could give us the detailed structure of a protein.

I was still, at this time, a beginning graduate student. By giving my colleagues a very necessary jolt I had deflected their attention in the right direction. In later years few people remembered this or appreciated my contribution except Bernal, who referred to it more than once. Of course in the long run my point of view was bound to emerge. All I did was to help create an atmosphere in which it happened a little sooner. I never wrote up my critique, though my notes for the talk survived for a few years. The main result as far as I was concerned was that Bragg came to regard me as a nuisance who didn't get on with experiments and talked too much and in too critical a manner. Fortunately he changed his mind later on.

I was, incidentally, not alone on my opinion. In those days most of the other crystallographers believed that protein crystallography was hopeless, or likely to come to fruition only in the next century. In this they were carrying their pessimism too far. I at least had a close acquaintance with the subject and could see one possible method of solving the problem. It is interesting to note the curious mental attitude of scientists working on "hopeless" subjects. Contrary to what one might at first expect, they are all buoyed up by irrepressible optimism. I believe there is a simple explanation for this. Anyone without such optimism simply leaves the field and takes up some other line of work. Only the optimists remain. So one has the curious phenomenon that workers in subjects in which the prize is big but the prospects of success very small always appear very optimistic. And this in spite of the fact that, although plenty appears to be going on, they never seem to get appreciably nearer their goal. Parts of theoretical neurobiology seem to me to have exactly this character.

Fortunately, solving the structure of protein by X-ray diffraction was not as hopeless as it had seemed to some. In 1962 Max Perutz and John Kendrew shared the Nobel Prize for Chemistry for their work on the structures of hemoglobin and myoglobin respectively.

Jim Watson, Maurice Wilkins, and I shared the Nobel Prize for Medicine or Physiology in that same year. The citation reads: ". . . for their discoveries concerning the molecular structures of nucleic acid and its significance for information transfer in living material." Rosalind Franklin, who had done such good work on the X-ray diffraction patterns of DNA fibers, had died in 1958.

5

The α Helix

SIR LAWRENCE BRAGG was one of those scientists with a boyish enthusiasm for research, which he never lost. He was also a keen gardener. When he moved in 1954 from his large house and garden in West Road, Cambridge, to London, to head the Royal Institution in Albemarle Street, he lived in the official apartment at the top of the building. Missing his garden, he arranged that for one afternoon each week he would hire himself out as a gardener to an unknown lady living in The Boltons, a select inner-London suburb. He respectfully tipped his hat to her and told her his name was Willie. For several months all went well till one day a visitor, glancing out of the window, said to her hostess, "My dear, what *is* Sir Lawrence Bragg doing in your garden?" I can think of few other scientists of his distinction who would do something like this.

Bragg had a great gift for seeing problems in simple terms, realizing that many apparent complications might fall away if the basic underlying pattern could be discovered. It was thus not surprising that in 1950 he wanted to show that at least some stretches of the polypeptide chain in a protein folded up in a simple manner. This was not an entirely new approach. Bill Astbury, the crystallographer, had tried to interpret his X-ray diagrams of keratin (the protein of hair and fingernails) using molecular models with regular repeats. He had found two forms of these fiber diagrams, which he called α and β. His idea for the β structure was not far from the correct answer, but his suggestion for the α structure was completely wide of the mark. This was partly because he was a sloppy model builder and was not meticulous enough about the distances and angles involved and partly because the experimental evidence was misleading in a way that would have been difficult for him to foresee.

It was well known that any chain with identical repeating links that fold so that every link is folded in exactly the same way, and with the same relations with its close neighbors, will form a helix (sometimes incorrectly called a spiral by nonmathematicians). The extreme solutions—a straight line or a circle—are regarded mathematically as degenerate helices.

Bragg's initial training had been as a physicist, and much of his work on molecular structure had been on inorganic materials such as the silicates. He did not have a detailed familiarity with organic chemistry or the related physical chemistry, though naturally he understood the elements of both these subjects. He decided that a good approach would be to build regular models of the polypeptide backbone, ignoring the complexities of the various side-chains.

A polypeptide chain has as its backbone a regular sequence of atoms, with the repeat . . . CH-CO-NH . . . (where C stands for carbon, H for hydrogen, O for oxygen, and N for nitrogen). The actual way that atoms are linked together is shown in appendix A. To each CH is attached a small group of atoms—often called R by chemists, where "R" stands for "Residue." Here we shall call R the side-chain. We know now that there are just twenty different side-chains commonly found in proteins. For the smallest residue, glycine, R is just a hydrogen atom—hardly a chain at all. The next largest is called alanine and has a methyl group (CH_3) as its side-chain. The others are of various sizes. Some carry a positive electric charge, some a negative one, and some no charge at all. Most of them are fairly small. The largest two, tryptophan and arginine, have only eighteen atoms in their side-chains. The names of all twenty of them (but not their formulas) are listed in appendix B.

Such a polypeptide chain is built up by joining together little molecules called amino acids. (The details of the chemistry are given in appendix A.) When a protein is synthesized, the relevant amino acids are joined together, head to tail, with the elimination of water, forming a long string called the polypeptide chain. As I have explained, the exact *order* of the amino acids in a particular protein, which is dictated by its gene, determines its character.

The α Helix

What we need to know is how each particular polypeptide chain is folded in the three-dimensional structure of the protein and exactly how all the side-chains (some of which are somewhat flexible) are arranged in space, so that we can understand how the protein does its job. Bragg and others wanted to find out, by model building, whether the main polypeptide chain could take up one or more regular folds. There were hints, from Astbury's α and β X-ray patterns, that the chain might well do so.

They therefore worked solely with the polypeptide backbone and ignored its side-chains. It may not be obvious why models had to be built at all, since the simple chemical structure of a unit of the backbone was well established. All the bond distances were known and all the bond angles. However, there can be fairly free rotation about bonds called single bonds (but not, by contrast, about double bonds), and the exact configuration of the atoms in space depends on just how these angles of rotation are fixed. This usually depends on interactions between atoms a little distant from one another down the chain and there may be several plausible alternatives, especially if these connections are weak ones.

The reason for this flexibility may not be immediately obvious. An easy way to see this is to use your hand. Place one of your hands so that the fingers are all in one plane, with the thumb exactly at right angles to the index finger. You can still waggle your thumb while preserving this right angle, yet the three-dimensional shape of your hand is changing. (See figure 5.1.) This is so even though all the (nearest neighbor) distances are constant—the length of the thumb and of each finger—as are the angles between them. Only the so-called dihedral angle (between the plane of the four fingers and the plane containing your thumb and your index finger) is changing. An example of an "interaction at a little distance" just referred to would be the changing distance between your thumbnail and the nail of your little finger.

In the case of a chemical molecule, there must be interactions of some sort if the molecule has to take up a particular configuration. It was clear that the best way for a polypeptide chain to hold itself together is for it to form hydrogen bonds between certain atoms in its backbone. Hydrogen bonds are weak bonds. The energy is only a small multiple of the thermal energy (at room temperature), and so a single hydrogen bond is easily broken by the constant thermal agitation. This is partly why water is a fluid at normal temperatures

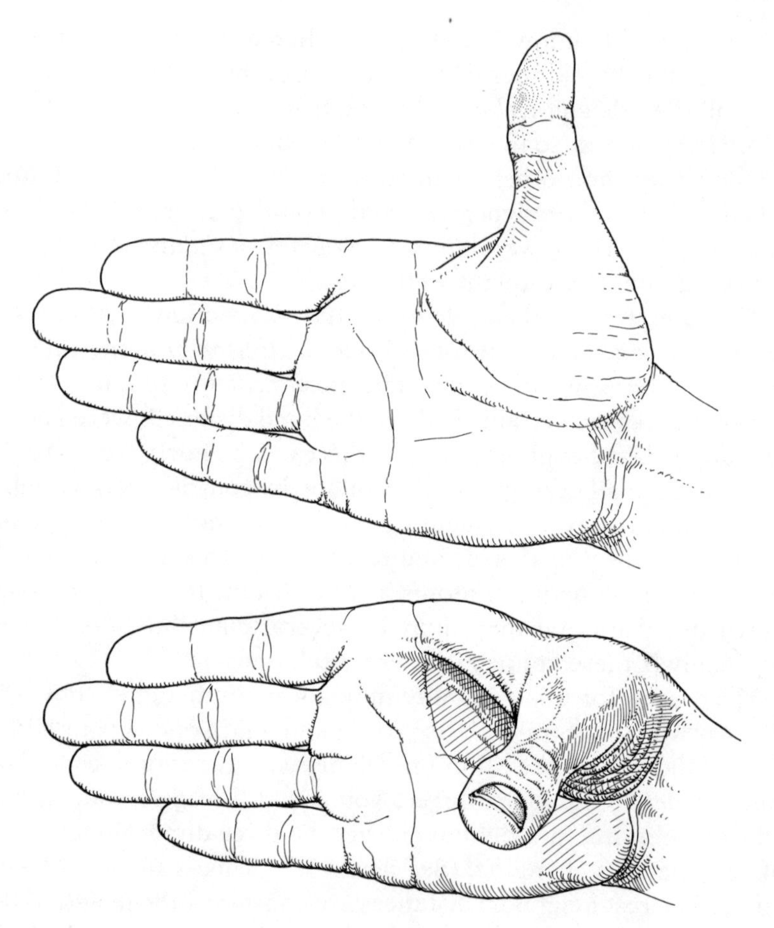

FIGURE 5.1
Showing how the thumb can be moved to give a different shape to the hand while preserving all the *direct* angles and distances.

and pressures. A hydrogen bond is formed from a donor atom (plus the hydrogen bonded to it) and a recipient. In a polypeptide chain the only strong donor is the NH group, and the only likely recipient the O of the CO group. John Kendrew pointed out that such a hydrogen bond in effect produces a particular ring of atoms. By enumerating all possible rings of this kind one can enumerate all possible structures of this type, each characterized by the NH group bonding to a particular CO group, say one that was three repeats away along the chain. This bonding is repeated over and over again down the length of the chain. The multiple hydrogen bonds thus formed help to stabilize the structure against the battering of thermal motion.

Using special model atoms, made of metal, and links built exactly to scale, Bragg, Kendrew, and Perutz systematically built all possible models, stopping only at folds that were not sufficiently compact. They hoped that one model would prove a much better fit to the X-ray data than all the others. Unfortunately they did not let the models take up their most favorable configurations. Astbury had shown that the α pattern had a strong X-ray spot on the so-called meridian, with a spacing corresponding to a repeat in the fiber direction of 5.1 Å. This implied that an important aspect of the structure repeated after this distance, probably the "pitch"—the distance between successive turns. Because this spot was exactly on the meridian, it suggested that the screw axis (the symmetry element associated with a regular helix) was an integer, though it did not say directly what the integer was. Bragg pointed out that it could be twofold, threefold, fourfold, or even fivefold or higher. As stated earlier, a wallpaper—a two-dimensional repeating pattern—cannot have fivefold symmetry, but there was no reason why a single polypeptide helix should not have a fivefold screw axis. This simply means that if you rotate the helix by 72 degrees (360 degrees divided by 5) and at the same time translate the structure along its axis a certain distance, it will look exactly the same, if you ignore any effects of the ends.

For this reason Bragg, Kendrew, and Perutz built all their models with integer axes. They also built them a little too sloppily. One particular group of atoms, the so-called peptide group, should really be planar—all the six atoms involved should be on or very close to one plane—whereas they allowed rotation about the peptide bond, which made their models too accommodating.

In short, they made one feature (the exact nature of the screw axis) too restrictive and they were too permissive about another—the planarity of the peptide bond. Not surprisingly, *all* their models looked ugly, and they were unable to decide which was best. Reluctantly they published their results in the *Proceedings of the Royal Society,* even though they were inconclusive. It so happened that I was asked to read the proofs of this paper (I believe the proofs were due to arrive when all three authors were away from the lab), but I was too ignorant of the fine points involved to see what was wrong.

Unbeknown to my colleagues, Linus Pauling was also following the same approach. He is now known to the general public mainly because of his championship of vitamin C. At that time he was probably the leading chemist in the world. He had pioneered the application of quantum mechanics to chemistry (explaining in the process, for example, why carbon has a valence of four) and was professor of chemistry at the California Institute of Technology, where he led several very talented groups of research workers. He was especially interested in using organic chemistry to explain important phenomena in biology.

Pauling has described how he first hit on the α helix while confined to bed with a cold during his stay in Oxford in 1948 as a visiting professor. His main paper on the α helix appeared, with several others of his works, in the *Proceedings of the National Academy of Sciences* in the spring of 1951. Pauling had known that the peptide bond was approximately planar, mainly because he had a more intimate acquaintance with organic physical chemistry than the three Cambridge workers. He had not attempted to make the structure with an integer screw but had let the models fold naturally into any screw they were comfortable with. The α helix turned out to have just 3.6 units per turn. He also noticed a paper by Bamford, Hanby, and Happey, the polymer workers, on the X-ray diffraction of a synthetic polypeptide that fit his model rather well. The fact that his model did not explain the 5.1 Å reflection on the meridian he put to one side. The irony was that Bragg, Kendrew, and Perutz had built, among other models, one that was, in effect, an α helix, but they had deformed the poor thing to make it have an exact fourfold axis. This made it look very forced, as indeed it was.

It soon became apparent that Pauling's α helix was the correct solution. Bragg was quite cast down. He walked slowly up the stairs. (When things went well for Rutherford he would bound upstairs singing "Onward Christian Soldiers.") "The biggest mistake of my scientific career," Bragg described it. The fact that it was Linus Pauling who had solved the problem didn't help, for Bragg had been beaten to the post before by Pauling. Perutz learned that after one of his own seminars a local physical chemist had told him that the peptide group ought to be planar. Perutz had even recorded it on his notes but had done nothing about it. It was not that they had not tried to get good advice, but some of what they had received had been unfortunate. Charles Coulsen, a theoretical chemist from Oxford, had told them, in my hearing, that the nitrogen atom might be "pyramidal," which was a highly misleading piece of information.

Honor was redeemed somewhat when Perutz spotted that the α helix should have a strong reflection on the meridian at 1.5 Å, corresponding to the height between successive stages of the helix, and duly found it. Together with two other crystallographers, Vladimir Vand at Glasgow University and Bill Cochran, in the Cavendish, I worked out the general nature of the Fourier Transform of a set of atoms arranged on a regular helix, and Cochran and I showed that it fit rather well the X-ray pattern of a synthetic polypeptide. But in some ways we were rubbing salt into our own wounds.

What, then, was the explanation of the misleading spot at 5.1 Å? A little later Pauling and I independently hit on the correct explanation. Because of their noninteger screw, α helices do not pack easily side by side. They pack best when there is a small angle between them, and, if they are deformed slightly, this leads to a coiled coil—that is, two or three α helices packed side by side but slowly coiling around one another [a nice example of symmetry breaking by a weak interaction]. This additional coiling threw the 5.4 Å off-meridianal spot onto the meridian at 5.1 Å.

It might be argued that since α helices are found almost exclusively in biological molecules, a model of a polypeptide backbone should not be rejected merely because it is ugly. I would prefer to say that because of its molecular simplicity the basic α helix is nearer to physical chemistry than to biology. At that level there are few alternatives for evolution to work on. It is only when we consider the side-chains, and the many ways a long polypeptide chain can fold back on itself, that a very large variety of structures become possible. Simplicity is then likely to yield to sophistication. Elegance, if it exists, may well be more subtle and what may at first sight seem contrived or even ugly may be the best solution that natural selection could devise.

This failure on the part of my colleagues to discover the α helix made a deep impression on Jim Watson and me. Because of it I argued that it was important not to place too much reliance on any single piece of experimental evidence. It might turn out to be misleading, as the 5.1 Å reflection undoubtedly was. Jim was a little more brash, stating that no *good* model ever accounted for *all* the facts, since some data was bound to be misleading if not plain wrong. A theory that *did* fit all the data would have been "carpentered" to do this and would thus be open to suspicion.

People have sometimes stated that Pauling's model of the α helix or his incorrect model for DNA gave us the idea that DNA was a helix. Nothing could be farther from the truth. Helices were in the air, and you would have to be either obtuse or very obstinate not to think along helical lines. What Pauling did show us was that exact and careful model building could embody constraints that the final answer had in any case to satisfy. Sometimes this could lead to the correct structure, using only a minimum of the direct experimental evidence. This was the lesson that we learned and that Rosalind Franklin and Maurice Wilkins failed to appreciate in attempting to solve the structure of DNA. That, and the necessity for making no assumptions that could not be doubted from time to time. It should also be said that Jim and I were highly motivated to succeed, even if we approached problems in a relaxed manner, were quick to spot success when we saw it and to learn what lessons we could draw both from successes and from failures.

The α helix was an important milestone on the rocky path of molecular biology but it did not have the same impact as the DNA double helix did. We initially hoped that, given the basic folds of the α helix and β sheets, we might be able to solve the structure of a protein by straightforward model building. Unfortunately most proteins are too complex and too sophisticated for that. In short, these two structural clichés alerted us to what to expect in some parts of a protein but did not immediately reveal the secret of the specificity and catalytic activity of a particular protein. The structure of DNA, on the other hand, immediately gave the game away, suggesting only too vividly how nucleic acid could be replicated exactly. DNA is, at bottom, a much less sophisticated molecule than a highly evolved protein and for this reason reveals its secrets more easily. We were not to know this in advance—it was just good luck that we stumbled onto such a beautiful structure.

Pauling was a more important figure in molecular biology than is sometimes realized. Not only did he make certain key discoveries (that sickle cell anemia is a molecular disease, for example), but he had the correct theoretical approach to these biological problems. He believed that much that we needed to explain could be done using the well-established ideas of chemistry and, in particular, the chemistry of macromolecules and that our knowledge of the various kinds of atoms, especially carbon, and of the bonds that hold atoms together [the homopolar bond, electrostatic interactions, hydrogen bonds, and van der Waal's forces] would be enough to crack the mysteries of life.

Max Delbrück, on the other hand, who started as a physicist, hoped that biology would enable us to discover new laws of physics. Delbrück also worked at Cal Tech, where Pauling was. He had pioneered important studies of certain viruses, called bacteriophage ("phage" for short), and was one of the leaders of the very influential Phage Group, of which Jim Watson was a more junior member. I don't think Delbrück much cared for chemistry. Like most physicists, he regarded chemistry as a rather trivial application of quantum mechanics. He had not fully imagined what remarkable structures can be built by natural selection, nor just how many distinct types of proteins there might be.

Time has shown that, so far, Pauling was right and Delbrück was wrong, as indeed Delbrück acknowledged in his book, *Mind into Matter.* Everything we know about molecular biology appears to be explainable in a standard chemical way. We also now appreciate that molecular biology is not a trivial aspect of biological systems. It is at the heart of the matter. Almost all aspects of life are engineered at the molecular level, and without understanding molecules we can only have a very sketchy understanding of life itself. All approaches at a higher level are suspect until confirmed at the molecular level.

6

How to Live with a Golden Helix

THE DOUBLE HELIX is indeed a remarkable molecule. Modern man is perhaps 50,000 years old, civilization has existed for scarcely 10,000 years, and the United States for only just over 200 years; but DNA and RNA have been around for at least several billion years. All that time the double helix has been there, and active, and yet we are the first creatures on Earth to become aware of its existence.

So much has already been written about our discovery of the double helix that it is difficult for me to add much to what has already been said. "Every schoolboy knows" that DNA is a very long chemical message written in a four-letter language. The backbone of each chain is almost entirely uniform. The four letters—the bases—are joined to the backbone at regular intervals. Normally the structure consists of two separate chains, wound around one another to form the double helix, but the helix is not the real secret of the structure. That lies in the way the bases are paired: adenine pairing with thymine, guanine with cytosine. In shorthand, A=T, G≡C, each dash representing a weak chemical bond, the hydrogen bond. It is this specific pairing between bases on opposite strands that is the heart of the replication process. Whatever sequence is written on one of the chains, the other chain must have the complementary sequence, given by the pairing rules. Biochemistry is mainly based on organic chemical molecules fitting closely together. DNA is no exception. (See appendix A for a slightly more detailed account.)

DNA was not always a familiar term, but even thirty years ago it was not entirely unknown. The physical chemist Paul Doty told me that shortly after lapel buttons came into style he was in New York and to his astonishment saw one with "DNA" written on it. Thinking it must refer to something else, he asked the vendor what it meant. "Get with it, Bud," the man replied in a strong New York accent. "Dat's the gene."

Nowadays most people know what DNA is, or if they don't they know it must be a dirty word, like "chemical" or "synthetic." Fortunately people who do recall that there are two characters called Watson and Crick are often not sure which is which. Many's the time I've been told by an enthusiastic admirer how much they enjoyed my book—meaning, of course, Jim's. By now I've learned that it's better not to try to explain. An even odder incident happened when Jim came back to work at Cambridge in 1955. I was going into the Cavendish one day and found myself walking with Neville Mott, the new Cavendish professor (Bragg had gone on to the Royal Institution in London). "I'd like to introduce you to Watson," I said, "since he's working in your lab." He looked at me in surprise. "Watson?" he said. "Watson? I thought your name was Watson-Crick."

Some people still find DNA hard to understand. I recall a singer in a nightclub in Honolulu telling me how, when she was a schoolgirl, she had cursed Watson and me because of the difficult things about DNA she had to learn in biology classes. Really the ideas needed to grasp the structure are, if properly presented, ridiculously easy, since they do not violate common sense, as quantum mechanics and relativity do. I believe there is a good reason for the simplicity of the nucleic acids. They probably go back to the origin of life, or at least very close to it. At that time mechanisms had to be fairly simple or life could not have started. Of course the very existence of chemical molecules can only be explained by quantum mechanics, but fortunately the shape of a chemical molecule can be embodied rather easily in a mechanical model, and it is this that makes the ideas easy to understand.

For those who have not already heard how the double helix was discovered, the following brief outline may help. Astbury, at Leeds University, had taken some poor but suggestive X-ray diffraction photographs of DNA fibers. After the Second World War Maurice Wilkins, working in Randall's laboratory at King's College, London, had obtained some rather better ones. Randall then hired an experienced crystallographer, Rosalind Franklin, to help solve the structure. Unfortunately Rosalind and Maurice found it difficult to work together. He wanted her to pay more attention to the wetter form (the so-called B form), which gave a simpler X-ray pattern but a more revealing one than that given by the slightly drier form (the A form), though the latter gave more detailed X-ray pictures.

At Cambridge I was working on a Ph.D. thesis about the X-ray diffraction of proteins. Jim Watson, a visiting American, then age twenty-three, was determined to discover what genes were and hoped that solving the structure of DNA might help. We urged the London workers to build models, using the approach Linus Pauling had used to solve the α helix. We ourselves produced a totally incorrect model, as did Linus Pauling a little later. Finally, after many ups and downs, Jim and I guessed the correct structure, using some of the experimental data of the London group together with Chargaff's rules about the relative amounts of the four bases in different sorts of DNA.

I first heard of Jim from Odile. One day when I came home she said to me, "Max was here with a young American he wanted you to meet and—you know what—he had no hair!" By this she meant that Jim had a crew cut, then a novelty in Cambridge. As time went on Jim's hair got longer and longer, as he tried to take on the local coloration, though he never got so far as to sport the long hair that men wore in the sixties.

Jim and I hit it off immediately, partly because our interests were astonishingly similar and partly, I suspect, because a certain youthful arrogance, a ruthlessness, and an impatience with sloppy thinking came naturally to both of us. Jim was distinctly more outspoken than I was, but our thought processes were fairly similar. What was different was our background knowledge. By that time I knew a fair amount about proteins and X-ray diffraction. Jim knew much less about these topics but a lot more about the experimental work on phages (bacterial viruses) and especially those associated with the Phage Group, led by Max Delbrück, Salva Luria, and Al Hershey. Jim also knew more about bacterial genetics. I suspect our knowledge of classical genetics was about the same.

Not surprisingly, we spent a lot of time talking over problems together. This did not pass unnoticed. Our group at the Cavendish had started with very little—for a brief period in 1949 we all worked in one room. By the time Jim joined us, Max and John Kendrew had a tiny private office. At this point the group was offered an extra room. It was not clear at first who should have this till one day Max and John, rubbing their hands together, announced that they were going to give it to Jim and me, ". . . so that you can talk to each other without disturbing the rest of us," they said. A fortunate decision, as it turned out.

When we met Jim had already obtained his doctorate, whereas I, though some twelve years older, was still a graduate student. Maurice Wilkins, in London, had done much of the initial X-ray work, which was then taken over and extended by Rosalind Franklin. Jim and I never did any experimental work on DNA, though we talked endlessly about the problem. Following Pauling's example, we believed the way to solve the structure was to build models. The London workers followed a more painstaking approach.

Our first attempt at a model was a fiasco, because I thought, quite erroneously, that the structure contained very little water. This mistake was partly due to ignorance on my part—I should have realized that a sodium ion was likely to be heavily hydrated—and partly due to Jim's misunderstanding of a technical crystallographic term that Rosalind had used in a seminar she gave. [He mixed up "asymmetric unit" and "unit cell."]

This was not our only mistake. Misled by the term tautomeric forms, I assumed that certain hydrogen atoms on the periphery of the bases could be in one of several positions. Eventually Jerry Donohue, an American crystallographer who shared an office with us, told us that some of the textbook formulas were erroneous and that each base occurred almost exclusively in one particular form. From that point on it was easy going.

The key discovery was Jim's determination of the exact nature of the two base pairs (A with T, G with C). He did this not by logic but by serendipity. [The logical approach—which we would certainly have used had it proved necessary—would have been: first, to assume Chargaff's rules were correct and thus consider only the pairs suggested by these rules, and second, to look for the dyadic symmetry suggested by the C2 space group shown by the fiber patterns. This would have led to the correct base pairs in a very short time.] In a sense Jim's discovery was luck, but then most discoveries have an element of luck in them. The more important point is that Jim was looking for something significant and *immediately recognized the significance of the correct pairs when he hit upon them by chance*—"chance favors the prepared mind." This episode also demonstrates that play is often important in research.

During the spring and summer of 1953 Jim Watson and I wrote four papers on the structure and function of DNA. The first appeared in *Nature* on April 25 accompanied by two papers from King's College, London, the first by Wilkins, Stokes, and Wilson, the other by Franklin and Gosling. Five weeks later we published a second paper in *Nature,* this time on the genetic implications of the structure. (The order of the authors' names on this paper was decided by the toss of a coin.) A general discussion was included in the volume that came from that year's Cold Spring Harbor Symposium, the subject of which was viruses. We also published a detailed technical account of the structure, with rough coordinates, in an obscure journal in the middle of 1954.

The first *Nature* paper was both brief and restrained. Apart from the double helix itself, the only feature of the paper that has excited comment was the short sentence: "It has not escaped our notice that the specific pairing we have postulated immediately suggests a possible copying mechanism for the genetic material." This has been described as "coy," a word that few would normally associate with either of the authors, at least in their scientific work. In fact it was a compromise, reflecting a difference of opinion. I was keen that the paper should discuss the genetic implications. Jim was against it. He suffered from periodic fears that the structure might be wrong and that he had made an ass of himself. I yielded to his point of view but insisted that something be put in the paper, otherwise someone else would certainly write to make the suggestion, assuming we had been too blind to see it. In short, it was a claim to priority.

Why, then, did we change our minds and, within only a few weeks, write the more speculative paper of May 30? The main reason was that when we sent the first draft of our initial paper to King's College we had not yet seen the papers by the researchers there. Consequently we had little idea of how strongly their X-ray evidence supported our structure. Jim had seen the famous "helical" X-ray picture of the B form reproduced by Franklin and Gosling in their paper, but he certainly had not remembered enough details to construct the arguments about Bessel functions and distances that the experimentalists gave. I myself at that time had not seen the picture at all. Consequently we were mildly surprised to discover that they had got so far and delighted to see how well their evidence supported our idea. Thus emboldened, Jim was easily persuaded that we should write a second paper.

I think what needs to be emphasized about the discovery of the double helix is that the path to it was, scientifically speaking, fairly commonplace. What was important was not the way it was discovered but the object discovered—the structure of DNA itself. You can see this by comparing it with almost any other scientific discovery. Misleading data, false ideas, problems of personal interrelationships occur in much if not all scientific work. Consider, for example, the discovery of the basic structure of collagen, the major protein of tendons, cartilage, and other tissues. The basic fiber of collagen is made of *three* long chains wound around one another. Its discovery had all the elements that surrounded the discovery of the double helix. The characters were just as colorful and diverse. The facts were just as confused and the false solutions just as misleading. Competition and friendliness also played a part in the story. Yet nobody has written even one book about the race for the triple helix. This is surely because, in a very real sense, collagen is not as important a molecule as DNA.

Of course this depends to some extent on what you consider important. Before Alex Rich and I worked (quite by accident, incidentally) on collagen, we tended to be rather patronizing about it. "After all," we said, "there's no collagen in plants." In 1955, after we got interested in the molecule, we found ourselves saying, "Do you realize that one-third of all the protein in your body is collagen?" But however you look at it, DNA *is* more important than collagen, more central to biology, and more significant for further research. So, as I have said before: It is the molecule that has the glamour, not the scientists.

One of the oddities of the whole episode is that neither Jim nor I were officially working on DNA at all. I was trying to write a thesis on the X-ray diffraction of polypeptides and proteins, while Jim had ostensibly come to Cambridge to help John Kendrew crystallize myoglobin. As a friend of Maurice Wilkins I had learned a lot about their work on DNA—which *was* officially recognized—while Jim had become intrigued by the diffraction problem after hearing Maurice talk in Naples.

People often ask how long Jim and I worked on DNA. This rather depends on what one means by work. Over a period of almost two years we often discussed the problem, either in the laboratory or in our daily lunchtime walk around the Backs (the college gardens that border the river) or at home, since Jim occasionally dropped in near dinnertime, with a hungry look in his eye. Sometimes, when the summer weather was particularly tempting, we would take the afternoon off and punt up the river toward Grantchester. We both believed that DNA was important though I don't think we realized just how important it would turn out to be. Originally my view was that solving the X-ray diffraction patterns of the DNA fibers was a job for Maurice and Rosalind and their colleagues at King's College, London, but as time went on both Jim and I became impatient with their slow progress and their pedestrian methods. The coolness between Rosalind and Maurice did not help matters.

The main difference of approach was that Jim and I had an intimate knowledge of the way the α helix was discovered. We appreciated what a strong set of constraints the known interatomic distances and angles provided and how postulating that the structure was a regular helix reduced the number of free parameters drastically. The King's workers were reluctant to be converted to such an approach. Rosalind, in particular, wanted to use her experimental data as fully as possible. I think she thought that to guess the structure by trying various models, using a minimum of experimental facts, was too flashy.

People have discussed the handicap that Rosalind suffered in being both a scientist and a woman. Undoubtedly there were irritating restrictions—she was not allowed to have coffee in one of the faculty rooms reserved for men only—but these were mainly trivial, or so it seemed to me at the time. As far as I could see her colleagues treated men and women scientists alike. There were other women in Randall's group—Pauline Cowan (now Harrison), for example—and moreover, their scientific advisor was Honor B. Fell, a distinguished tissue culturist. The only opposition I ever heard about was that of Rosalind's family. She came from a solid banking family who felt that a nice Jewish girl should get married and have babies, rather than devote her life to scientific research, but even they did not provide really active opposition to her choice of a career.

But, in spite of her freedom to pursue research as she wished, I think there were more subtle handicaps. Part of the problem Rosalind had with Maurice was her suspicions that he really wanted her as an assistant rather than as an independent worker. Rosalind did not herself choose to work on DNA because she thought it to be biologically important. When John Randall first offered her a job, it was so that she could study the X-ray diffraction of proteins in solution. Rosalind's previous work on the X-ray diffraction of coal was well suited as an introduction to such a study. Then Randall changed his mind and suggested that, as the DNA fiber work (which Maurice had been doing) had become interesting, it might be better if she worked on that. I doubt if Rosalind knew very much about DNA before Randall suggested that she work on it.

Feminists have sometimes tried to make out that Rosalind was an early martyr to their cause, but I do not believe the facts support this interpretation. Aaron Klug, who knew Rosalind well, once remarked to me, with reference to a book by a feminist, that "Rosalind would have hated it." I don't think Rosalind saw herself as a crusader or a pioneer. I think she just wanted to be treated as a serious scientist.

In any event, Rosalind's experimental work was first class. It is difficult to see how it could be bettered. She was less at home, however, in the detailed interpretation of the X-ray photographs. Everything she did was sound enough—almost too sound. She lacked Pauling's panache. And I believe that one reason for this, apart from the marked difference in temperament, was because she felt that a woman must show herself to be fully professional. Jim had no such anxieties about his abilities. He just wanted the answer, and whether he got it by sound methods or flashy ones did not bother him one bit. All he wanted was to get it *as quickly as possible.* People have argued that this was because we were overcompetitive, but the facts hardly support this. In our enthusiasm for the model-building approach we not only lectured Maurice on how to go about it but even lent him our jigs for making the necessary parts of the model. In some ways I can see that we behaved insufferably (they never did use our jigs), but it was not all due to competitiveness. It was because we passionately wanted to know the details of the structure.

This, then, was a powerful force in our favor. I believe there were at least two others. Neither Jim nor I felt any external pressure to get on with the problem. This meant that we could approach it intensively for a period and then leave it alone for a bit. Our other advantage was that we had evolved unstated but fruitful methods of collaboration, something that was quite missing in the London group. If either of us suggested a new idea the other, while taking it seriously, would attempt to demolish it in a candid but nonhostile manner. This turned out to be quite crucial.

In solving scientific problems of this type, it is almost impossible to avoid falling into error. I've already listed some of my mistaken ideas. Now, to obtain the correct solution of a problem, unless it is transparently easy, usually requires a *sequence* of logical steps. If one of these is a mistake, the answer is often hidden, since the error usually puts one on completely the wrong track. It is therefore extremely important not to be trapped by one's mistaken ideas. The advantage of intellectual collaboration is that it helps jolt one out of false assumptions. A typical example was Jim's initial insistence that the phosphates must be on the inside of the structure. His argument was that the long basic amino acids of the histones and protamines (proteins associated with DNA) could then reach into the structure to contact the acidic phosphate groups. I argued at length that this was a very feeble reason and that we should ignore it. "Why not," I said to Jim one evening, "build models with the phosphates on the outside?" "Because," he said, "that would be too easy" (meaning that there were too many models he could build in this way). "Then why not try it?" I said, as Jim went up the steps into the night. Meaning that so far we had not been able to build even one satisfactory model, so that even one acceptable model would be an advance, even if it turned out not to be unique.

This argument had the important effect of directing our attention to the bases. While the phosphates were inside the structure, with the bases on the outside, we could afford to ignore the shape and position of the bases. As soon as we wanted to put them inside, we were forced to look at them more closely. I was amused to discover, when we finally built the bases to scale, that they differed in size from my previous mental picture of them—they were distinctly bigger—though their shape was close to the pictures in my mind.

There is thus no straightforward answer to the question of how long it took us. We had one intensive period of model building toward the end of 1951 but after that I myself was forbidden, for a period, to do anything further, as I was still a graduate student. For a week or so in the summer of 1952 I had experimented to see if I could find evidence for bases pairing in solution, but the necessity of working on my thesis made me abandon this approach too soon. The final attack, including the measurements of our model's coordinates, only took a few weeks. Hardly more than a month or so after that our papers appeared in *Nature.* It seems a ridiculously short period of work but all the hours and hours of reading and discussion that led up to the final model really should be included.

It soon transpired that our model was not even correct in detail. We had only two hydrogen bonds in our G=C pair, though we recognized that there might be three. Pauling subsequently made a decisive argument for three and was rather cross when the illustration in my *Scientific American* article showed only two. This, as it happened, was not really my fault, as the editor was in such a hurry (as is usually the case) that I never saw the proofs of the diagrams. We had also put the bases too far from the axis of the structure, but these errors did not alter the fact that our model captured all the essential aspects of the double helix. The two helical chains, running antiparallel, a feature I had deduced from Rosalind's own data; the backbone on the outside, with the bases stacked on the inside; and, above all, the key feature of the structure, the specific pairing of the bases.

Certain points are sometimes overlooked. It took courage (or rashness, according to your point of view) and a degree of technical expertise to put firmly to one side the difficult problem of unwinding the double helix and to reject a side-by-side structure. Such a model was suggested by the cosmologist George Gamow not long after ours was published, and it has been suggested again more recently by two other groups of authors. Let me skip forward in time to discuss these two models. In both of them the two DNA chains were not intertwined, as in ours, but lay side by side. This, they argued, would make it easier for the chains to separate during replication. Each chain did a sort of shimmy so that, at a first quick glance, the proposed configurations didn't look unlike our own. They claimed that these new models fit the X-ray data at least as well as ours did, if not better.

I didn't believe a word of this. I doubted very much the claims about the diffraction pattern, since such models would be expected to produce at least a few spots in those characteristic empty spaces in the X-ray fiber diagrams that a true helix produces. Moreover the models were ugly in that the shapes they took were forced on them by the model builders and seemed to exist for no obvious structural reason.

Such arguments, however, are not decisive and could easily be attributed to mere prejudice on my part. The two groups of innovators felt rather acutely that they were on the fringes of the scientific world. They feared that The Establishment would not listen to them. Quite the contrary was the case since everyone, including the editor of *Nature,* was bending over backward to give them a fair hearing.

At about this time Bill Pohl, a pure mathematician, got into the act. He pointed out, quite correctly, that unless something very special happened, the most likely result of replicating a piece of *circular* DNA would be two *interlocked* daughter circles rather than two separate ones. From this he deduced that the DNA chains could not be intertwined, as we had suggested, but had to lie side by side.

I corresponded at some length with him as well as talking to him on the phone. Later on he paid me a visit. He had become very well informed about experimental details and persisted strongly in his view. I told him in a letter that if nature did occasionally produce two interlinked circles, a special mechanism would have been evolved to unlink them. I believe he thought this an outrageous example of special pleading and was not at all convinced by it. It turned out, some years later, that this is exactly what does happen. Nick Cozzarelli and his co-workers showed that a special enzyme, called topoisomerase II, can cut both strands of a piece of DNA, pass another piece of DNA between the two ends, and then join the broken ends together again. It can thus unlink two linked DNA circles, and can even, at high enough concentrations of DNA, produce linked circles from separate ones.

Fortunately some brilliant work by Walter Keller and by Jim Wang on the "linking number" of circular DNA molecules proved that all these side-by-side models must be wrong. The two DNA chains in circular DNA were shown to wind around each other about the number of times our model predicted. I had spent so much time on this problem that in 1979 Jim Wang, Bill Bauer, and I wrote a review article "Is DNA Really a Double Helix?" setting out all the relevant arguments in some detail.

I doubt if even this, by itself, would convince a hardened skeptic, though at about that time Bill Pohl threw in the towel. Fortunately there was a new development. The reason a *decisive* argument could not be made from the previous X-ray data alone was partly that the X-ray photos didn't contain enough information and also because one had to *assume* a tentative model and then test it against the rather sparse data.

By the late 1970s the chemists had found an efficient way of synthesizing reasonable amounts of short stretches of DNA with any required base sequence. With luck, such a short stretch could be crystallized. Its structure could then be determined by X-ray diffraction, using unambiguous methods such as the isomorphous replacement method, which involved no prior assumptions about the result. Moreover the X-ray spots from such crystals extended to a much higher resolution than the old fiber diagrams did, partly because the fiber was produced from DNA that had all sorts of different sequences mixed together. Not surprisingly, fibers gave a more blurred picture of the molecule, since what the X rays see is the *average* structure of all the molecules.

The first result (around 1980) on these small bits of DNA, by Alex Rich and his group at M.I.T. and also by Dick Dickerson and his colleagues at Cal Tech, produced another surprise. The X rays showed a *left-handed* structure, never seen before, with a zigzag appearance. It was christened Z-DNA. Its X-ray pattern was quite unlike the classical DNA patterns, so that it was clearly a new form of DNA. It turns out that Z-DNA forms most easily only with a special type of base sequence (alternating purines and pyrimidines). Exactly what nature uses Z-DNA for is still a hot topic of research; it may well be used in control sequences.

More ordinary DNA sequences were soon crystallized. This time the resulting structures looked very like those predicted by the X-ray fiber data, though there were small modifications and the helix varied somewhat depending on the local sequence of the bases. This also is still being actively studied.

The double-helical structure of DNA was thus finally confirmed only in the early 1980s. It took over twenty-five years for our model of DNA to go from being only rather plausible, to being *very* plausible (as a result of the detailed work on DNA fibers), and from there to being virtually certainly correct. Even then it was correct only in outline, not in precise detail. Of course the fact that base sequences were complementary (the key to its function) and that the two chains run in opposite directions was firmly established somewhat earlier by the chemical and biochemical work on DNA sequences.

The establishment of the double helix could serve as a useful case history, showing one example of the complicated way theories become "fact." I suspect that after about twenty to twenty-five years many human beings have a desire to overturn the old orthodoxy. Each generation needs a new music. In the case of the double helix, the hard bite of scientific facts made the new models unacceptable. In nonscientific subjects it is more difficult to repel the challenge and often the new ideas take over, mainly because of their novelty. Freshness is all. In both cases the new approach tries to preserve some aspects of the older viewpoint, for innovation is most effective when it builds on at least part of the existing tradition.

What, then, do Jim Watson and I deserve credit for? If we deserve any credit at all, it is for persistence and the willingness to discard ideas when they became untenable. One reviewer thought that we couldn't have been very clever because we went on so many false trails, but that is the way discoveries are usually made. Most attempts fail not because of lack of brains but because the investigator gets stuck in a cul-de-sac or gives up too soon. We have also been criticized because we had not perfectly mastered all the very diverse fields of knowledge needed to guess the double helix, but at least we were *trying* to master them all, which is more than can be said for some of our critics.

However, I don't believe all this amounts to much. The major credit I think Jim and I deserve, considering how early we were in our research careers, is for selecting the right problem and sticking to it. It's true that by blundering about we stumbled on gold, but the fact remains that we were looking for gold. Both of us had decided, quite independently of each other, that the central problem in molecular biology was the chemical structure of the gene. The geneticist Hermann Muller had pointed this out as long ago as the early 1920s, and many others had done so since then. What both Jim and I sensed was that there might be a shortcut to the answer, that things might not be *quite* as complicated as they seemed. Curiously enough, I believed this partly because of my very detailed grasp of the current knowledge of proteins. We could not at all see what the answer was, but we considered it so important that we were determined to think about it long and hard, from any relevant point of view. Practically nobody else was prepared to make such an intellectual investment, since it involved not only studying genetics, biochemistry, chemistry, and physical chemistry (including X-ray diffraction—and who was prepared to learn that?) but also sorting out the essential gold from the dross. Such discussions, since they tend to go on interminably, are very demanding and sometimes intellectually exhausting. Nobody without an overwhelming interest in the problem could sustain them.

And yet history of other theoretical discoveries often shows exactly the same pattern. In the broad perspective of the exact sciences we were not thinking very hard, but we were thinking a lot harder than most people in that corner of biology, since in those days, except for geneticists and possibly the people in the Phage Group, most of biology was not thought of as having a highly structured logic.

Then there is the question of what would have happened if Watson and I had not put forward the DNA structure. This is "iffy" history, which I am told is not in good repute with historians, though if a historian cannot give plausible answers to such questions I do not see what historical analysis is about. If Jim had been killed by a tennis ball, I am reasonably sure I would not have solved the structure alone, but who would? Jim and I always thought that Linus Pauling would be bound to have another shot at the structure once he had seen the King's College X-ray data, but he has stated that even though he immediately liked our structure it took him a little time to decide finally that his own was wrong. Without our model he might never have done so. Rosalind Franklin was only two steps away from the solution. She needed to realize that the two chains must run in opposite directions and that the bases, in their correct tautomeric forms, were paired together. She was, however, on the point of leaving King's College and DNA, to work instead on Tobacco Mosaic Virus with Bernal. (She died five years later at the early age of thirty-seven.) Maurice Wilkins had announced to us, just before he knew of our structure, that he was going to work full time on the problem. Our persistent propaganda for model building had also had its effect, and he was proposing to give it a try. Had

Jim and I not succeeded, I doubt whether the discovery of the double helix could have been delayed for more than two or three years.

There is a more general argument, however, proposed by Gunther Stent and supported by such a sophisticated thinker as Peter Medawar. This is that if Watson and I had not discovered the structure, instead of being revealed with a flourish it would have trickled out and that its impact would have been far less. For this sort of reason Stent had argued that a scientific discovery is more akin to a work of art than is generally admitted. Style, he argues, is as important as content.

I am not completely convinced by this argument, at least in this case. Rather than believe that Watson and Crick made the DNA structure, I would rather stress that the structure made Watson and Crick. After all, I was almost totally unknown at the time, and Watson was regarded, in most circles, as too bright to be really sound. But what I think is overlooked in such arguments is the intrinsic beauty of the DNA double helix. It is the molecule that has style, quite as much as the scientists. The genetic code was not revealed all in one go, but it did not lack for impact once it had been pieced together. I doubt if it made all that much difference that it was Columbus who discovered America. What mattered much more was that people and money were available to exploit the discovery when it was made. It is this aspect of the history of the DNA structure that I think demands attention, rather than the personal elements in the act of discovery, however interesting they may be as an object lesson (good or bad) to other workers.

It is really for the historian of science to decide how our structure was received. This is not an easy question to answer because there was naturally a spectrum of opinion that changed with time. There is no doubt, however, that it had a considerable and immediate impact on an influential group of active scientists. Mainly due to Max Delbrück, copies of the initial three papers were distributed to all those attending the 1953 Cold Spring Harbor Symposium, and Watson's talk on DNA was added to the program. A little later I gave a lecture at the Rockefeller Institute in New York, which I am told produced considerable interest, partly I think because I mixed an enthusiastic presentation of our ideas with a fairly cool assessment of the experimental evidence, roughly on the lines of the article that appeared in *Scientific American* in October 1954. Sydney Brenner, who had just finished his Ph.D. at Oxford under Hinshelwood, appointed himself, in the summer of 1954, as our representative at Cold Spring Harbor. He took some pains to get the ideas over to Milislav Demerec, who was then the director. (Sydney was to move from South Africa to Cambridge in 1957. He became my closest colleague, sharing an office with me for almost twenty years.) But not everyone was convinced. Barry Commoner (now an environmentalist) insisted, with some force, that physicists oversimplified biology, in which he was not completely wrong. Chargaff, when I visited him in the winter of 1953–54, told me (with his customary insight) that while our first paper in *Nature* was interesting, our second paper on the genetic implications was no good at all. I was mildly surprised to find when, in 1959, I talked with Fritz Lipmann (the distinguished biochemist), who had arranged for my lecture series at the Rockefeller, I learned that he had not really grasped our scheme of DNA replication. (It emerged that he had been talking to Chargaff.) By the end of the lectures, however, he gave a remarkably clear outline of our ideas in his summing up. The biochemist Arthur Kornberg has told me that when he began work on DNA replication, he did not believe in our mechanism, but his own brilliant experiments soon made him a convert, though always a careful and critical one. His work produced the first good experimental evidence that the two chains run in opposite directions. All in all, it seems to me that we got a very fair hearing, better than Avery and certainly a lot better than Mendel.

What was it like to live with the double helix? I think we realized almost immediately that we had stumbled onto something important. According to Jim, I went into the Eagle, the pub across the road where we lunched every day, and told everyone that we'd discovered the secret of life. Of that I have no recollection, but I do recall going home and telling Odile that we seemed to have made a big discovery. Years later she told me that she hadn't believed a word of it. "You were always coming home and saying things like that," she said, "so naturally I thought nothing of it." Bragg was in bed with flu at the time, but as soon as he saw the model and grasped the basic idea he was immediately enthusiastic. All past differences were forgiven and he became one of our strongest supporters. We had a constant stream of visitors, a contingent from Oxford that included Sydney Brenner, so that Jim soon began to tire of my repetitious enthusiasm. In fact at times he had cold feet, thinking that perhaps it was all a pipe dream, but the experimental data from King's College, when we finally saw them, were a great encouragement. By summer most of our doubts had vanished and we were able to take a long cool look at the structure, sorting out its accidental features (which were somewhat inaccurate) from its really fundamental properties, which time has shown to be correct.

For a number of years after that, things were fairly quiet. I named my family's Cambridge house in Portugal Place "The Golden Helix" and eventually erected a simple brass helix on the front of it, though it was a single helix rather than a double one. It was supposed to symbolize not DNA but the basic idea of a helix. I called it golden in the same way that Apuleius called his story "The Golden Ass," meaning beautiful. People have often asked me whether I intend to gild it, but we never got further than painting it yellow.

Finally one should perhaps ask the personal question—am I glad that it happened as it did? I can only answer that I enjoyed every moment of it, the downs as well as the ups. It certainly helped me in my subsequent propaganda for the genetic code. But to convey my own feelings, I cannot do better than quote from a brilliant and perceptive lecture I heard years ago in Cambridge by the painter John Minton in which he said of his own artistic creations, "The important thing is to be there when the picture is painted." And this, it seems to me, is partly a matter of luck and partly good judgment, inspiration, and persistent application.

There was in the early fifties a small, somewhat exclusive biophysics club at Cambridge, called the Hardy Club, named after a Cambridge zoologist of a previous generation who had turned physical chemist. The list of those early members now has an illustrious ring, replete with Nobel laureates and Fellows of the Royal Society, but in those days we were all fairly young and most of us not particularly well known. We boasted only one F.R.S.—Alan Hodgkin—and one member of the House of Lords—Victor Rothschild. Jim was asked to give an evening talk to this select gathering. The speaker was customarily given dinner first at Peterhouse. The food there was always good but the speaker was also plied with sherry before dinner, wine with it, and, if he was so rash as to accept them, drinks after dinner as well. I have seen more than one speaker struggling to find his way into his topic through a haze of alcohol. Jim was no exception. In spite of it all he managed to give a fairly adequate description of the main points of the structure and the evidence supporting it, but when he came to sum up he was quite overcome and at a loss for words. He gazed at the model, slightly bleary-eyed. All he could manage to say was "It's so beautiful, you see, so beautiful!" But then, of course, it was.

7

Books and Movies About DNA

OVER THE YEARS the discovery of the double helix has attracted the attention of a wide variety of people, from historians of science to Hollywood filmmakers. The best-known written account is Jim Watson's *The Double Helix.* This was a best-seller when first published in 1968 and has sold steadily ever since. It attracted a lot of interesting reviews, the best of which are included in the critical edition, published by Norton. Chargaff, rather typically, refused to allow his own review to be reprinted. There is an excellent review of the reviews by Gunther Stent, which puts the book and the various reviewers firmly and accurately into perspective.

I recall that when Jim was writing his book he read a chapter to me while we were dining together at a small restaurant near Harvard Square. I found it difficult to take his account seriously. "Who," I asked myself, "could possibly want to read stuff like this?" Little did I know! My years of concentration on the fascinating problems of molecular biology had, in some respects, led me to live in an ivory tower. Since all the people I met were mainly concerned with the intellectual interest of these problems, I must have tacitly assumed that everyone was like that. Now I know better. The average adult can usually enjoy something only if it relates to what he knows already, and what he knows about science is in many cases pitifully inadequate. What almost everybody is familiar with is the vagaries of personal behavior. People find it much easier to appreciate stories of competition, frustration, and animosity, against a background of parties, foreign girls, and punting on the river, than the details of the science involved.

I now appreciate how skillful Jim was, not only in making the book read like a detective story (several people have told me they were unable to put it down) but also by managing to include a surprisingly large amount of the science, although naturally the more mathematical parts had to be left out. The only surprising part of the book is Jim's reference to his thinking about the Nobel Prize. Max Perutz, John Kendrew, and I had never heard Jim talk in this way, so that if he really was thinking about Stockholm he must have kept it strictly to himself. To us he appeared strongly motivated by the scientific importance of the problem. It didn't occur to me that our discovery was prizeworthy till as late as 1956, and then only because of a casual remark Frank Putnam made to me on the subject.

Fortunately for those who really want to know what it was all about, more scholarly works exist. Robert Olby, in *The Path to the Double Helix,* has taken the story from the development of the idea of macromolecules up to the discovery itself. Horace Freeland Judson's account, entitled *The Eighth Day of Creation* (probably suggested by the publisher), is in some ways more vivid, since it contains lengthy verbatim quotations from most of the participants. His story begins nearer in time to the discovery of the double helix and continues for another dozen years or so until the genetic code was unraveled. Both are big, thick books. They may take a little time to get into but they provide the most complete and the most balanced accounts so far of the beginnings of classical molecular biology.

In the early 1970s I was approached by the late Ronnie Fouracre, who wanted to make a documentary about the discovery. Jim and Maurice agreed to take part. The shooting at Cambridge took about three days, a small part of it being shot in the Eagle. Afterward Odile and I hosted a lively party for the film crew at the Golden Helix—so lively that Ronnie regretted he hadn't brought his cameras along to shoot some of it for the film. The filming itself was strenuous but enjoyable. Only when it was all over did I realize that in the excitement I had completely forgotten Odile's birthday, something I have never done before or since.

Ronnie made two distinct versions. One, a more technical film, was for universities and schools. The other was for a lay audience. He had some trouble getting the latter film into shape and about three distinct versions were produced, partly in collaboration with the BBC. I thought the final version, with the commentary by Isaac Asimov, was the best. One version or another appeared under the Horizon label in England or the Nova label in the States.

Over the years there were jokes about other possible formats. Could it be made into a musical comedy, for example? Sydney Brenner had worked out a scenario of the story as a Western. Jim was to be the lone cowboy; Max, the telegraph clerk; and I, the riverboat gambler! The details, lovingly embroidered, produced much hilarity in his listeners.

Jim had other ambitions. He hoped for a full-length feature movie. From 1976 I lived in Southern California and occasionally met people from the film world. At one point 20th Century Fox appeared to show some interest, but they did not follow up. Eventually we were approached by Larry Bachmann, a well-established American film producer. I was very reluctant to give my permission. Larry allowed Odile and me, with two friends, to see part of the shooting of his latest film *Whose Life Is It, Anyway?* Later he asked us to see the "rough"—the first complete, though in some respects unfinished, version.

Before going to Hollywood, I had decided to oppose the making of any movie about our discovery of the double helix, and had even drafted a letter to that effect, but seeing the film Larry had produced made me change my mind. He had managed to handle an important theme in a serious manner, relieved by many light touches of humor. Before long Jim and I had acquired both a Hollywood agent and a Hollywood lawyer. We visited a couple of other producers who had expressed an interest but they seemed to us caricatures of the "typical" Hollywood producer, being mainly concerned with turning the story into another blood-and-thunder. Larry, on the other hand, showed a serious interest in the discovery, though what mainly appealed to him was the drama of the story and the cast of characters. And what a cast! The Brash Young Man from the Midwest, the Englishman who talks too much (and therefore must be a genius since geniuses either talk all the time or say nothing at all), the older generation, replete with Nobel Prizes, and best of all, a Liberated Woman who appears to be unfairly treated. And in addition some of the characters actually quarrel, in fact almost come to blows. The layman is delighted to learn that after all, in spite of science being so impossibly difficult to understand, SCIENTISTS ARE HUMAN, even though the word human more accurately describes the behavior of mammals rather than anything peculiar to our own species, such as mathematics.

Larry took some pains to read up on the various accounts of the discovery and to talk to many of the people involved. Before he could begin, a long contract that dealt with all foreseeable contingencies had to be written and signed. For example, it was set out at length exactly what share of the profits (if any) we would get if indeed a musical comedy were made. We also, as I recall, retained any comic book rights. We obtained these concessions because no filmmaker likes to start on a film about someone still living unless that person has signed a release that an actor may impersonate him. Otherwise there is the danger that the filmmaker may have a legal injunction slapped on him in the middle of shooting, which financially would be ruinous, whatever the outcome. We had some minor degree of protection: We could sue them if they imputed criminal acts to us, or acts of sexual perversion, but if they damaged our professional reputation we were to have no recourse. We soon learned that, as in other walks of life, the man who pays the piper calls the tune. It may take a quarter of a million dollars to produce a screenplay, while the whole movie is likely to cost something on the order of ten million dollars. The more money involved, the less say one has. "I hope you realize," said our agent, the first time we met him, "that they can put in anything about you they like." When we taxed Larry with this he simply said, "You have to trust me," and up to a point we did.

However, I told Larry that I believed that it was impossible to make a full-length feature movie of the story, since it did not contain enough sex and violence. Over a period of several years he and various coauthors tried hard to produce a suitable film script, but eventually it turned out as I predicted. The final version was rejected by the backers, even though a small amount of violence and sex had been added to color up the story.

It must be a general rule that the more sophisticated the treatment of a story, the smaller the audience it can command. The audience needed to make a feature movie pay is far too big for the DNA story. Rather, the story is more suitable for a play or, possibly, a limited-distribution movie. The problem is not helped by the fact that the older members of the potential audience, although they may have heard of DNA, hardly know what it is, whereas to some of the younger members the structure is old hat, since they learned all about it at school.

Larry Bachmann now lives in a charming manor house in a village a few miles from Oxford. He was given dining rights at Green College, and they liked him so much that they made him a Fellow. He keeps himself busy reorganizing Oxford tennis (being a keen tennis player), encouraging the local theatricals, and even advising the university on how to raise money. We meet from time to time, either at Oxford or at the Beverly Hills Tennis Club, to chat about the ways of the world.

In 1984 Jim and I were approached by the BBC. Mick Jackson, a BBC producer, wanted to make a docudrama about the discovery of DNA. ("Docudrama" implies something between a documentary and a drama.) It would attempt to be closer to the facts than the usual movie treatment but would shape the story to make it theatrically attractive. Jim and I and the other characters would be played by actors.

I was in favor of the BBC doing something, mainly because of its reputation for careful and fairly accurate productions. Jim, although at first attracted, eventually withdrew his collaboration, saying to me that he thought the BBC treatment would be too dull. What exactly Jim had in mind for a more exciting version was never spelled out.

I was consulted by both Mick Jackson and the scriptwriter, Bill Nicholson. Much research was done by Jane Callender, who became very familiar with the characters involved and the details of the story. The 106-minute program, called *Life Story,* went on the air in England on April 27, 1987. The American version, called *Double Helix,* went out on the Arts and Entertainment channel later that year. Jim is played by Jeff Goldblum, I am played by Tim Pigott-Smith, Maurice by Alan Howard, and Rosalind by Juliet Stevenson. Most of the reviews were favorable, as was the considerable phone-in response to the BBC. I was mildly surprised that it had been so well received, but Mick told me that a large segment of the British viewing public were astonished to find that scientists behaved as human beings. When I said that I thought Jim's book had already made that idea very familiar, Mick pointed out that many TV viewers had probably never read it.

The program closely follows the main lines of the story. It shows Rosalind in Paris, with her friend and scientific advisor Vittorio Luzzati, before she moved to King's College, London, to work on DNA in John Randall's lab there. It rather overemphasizes the differences Rosalind found, as a woman, between Paris and London. Maurice and Rosalind's failure to collaborate is brought out clearly. At Cambridge we see Jim being introduced to the college scene by Max Perutz, then meeting me. The fiasco of our first model-building attempt and the reactions of the King's workers are clearly delineated, though our telling-off by Bragg is quite fictitious. Other scenes include our meeting with Chargaff and our discussion with John Griffith about base pairing. Brash young Peter Pauling, Linus's son, is seen arriving in Cambridge. A little later he produces a copy of his father's scientific paper with the incorrect three-chain model of DNA. Rosalind loses her temper with Jim when he comes to London to show her Linus's paper. Maurice, in sympathy, shows Jim the revealing photograph of the B form, which Rosalind had taken but had put aside while she plodded on with the more detailed photos of the A form. Viewers had previously been prepared, so that they would appreciate the significance of this photo, by a little lecture I give to Jim on the diffraction of X rays by a helix. There is no doubt that seeing this dramatic photo prodded us into action, but in fact much of its data were made available to us in other ways.

Finally we see Jerry Donohue telling us that we had the wrong formulas [tautomeric forms] for the bases, so that Jim was able to hit on the correct base pairs. After that the model was almost inevitable. We see a very hyped-up version of this climax, followed by a stream of visitors, while the model of the double helix appears to rotate to celestial music. The film ends with Rosalind viewing the model and Jim chatting to his sister on a bridge over the Cam.

It is difficult for me to pass judgment on *Life Story* because I had been so close to the actual events. Almost everyone enjoys watching the tale as it unfolds on the screen. In spite of the intention to soft-pedal the science, a surprisingly large amount has been included, though I doubt if most viewers realize that DNA is not a short fat molecule but a long thin one. If we had made our model a more typical length it would have reached well above the clouds. The model we built was only a tiny fraction of the sort of lengths found in nature.

It is obviously unfair to criticize the BBC for not achieving complete factual accuracy. Anyone interested in knowing what really happened will get much closer to the truth by reading the printed accounts described earlier. What *Life Story* was trying to do was to get over the general nature of the discovery and to show in broad terms how it was done and how it was received.

The BBC, while striving to be factually correct, had no qualms in telescoping incidents and shifting scenes. The conversation among Maurice, Jim, and myself, shown as taking place in the college gardens near the river, actually occurred in my dining room at home. The party, with the men dressed as clergymen, in reality took place at Peter Mitchell's house; but the conversation between John Griffith and myself at that party occurred in a quiet pub. Nor did we meet Chargaff over a college dinner. But these translocations seem to me to be perfectly acceptable, since they get over important parts of the story and also the local atmosphere, even if the combinations shown were not the real ones.

There are a few more significant errors. Surprising as it may seem, I don't believe that Chargaff's rules were in the forefront of Jim's mind when he first stumbled on the correct base pairs. A more serious mistake are the words put into Rosalind's mouth. She says to Maurice Wilkins, "But you may be guessing right or you may not. We won't know until we've done the work. When we've done the work we won't need the guesses because we'll know the answer. [So] what's the point of the guesses?"

This argument appears, on the surface, to have considerable force, but it is incorrect. As explained earlier, the X rays provide only half the required data. For this reason a good model is worth its weight in gold, especially if, as in the DNA case, the X-ray reflections are rather few. Rosalind is unlikely to have said such words. If she had, it would have shown that she had not adequately grasped the problem confronting her.

It is implied, but not actually stated, that Rosalind and her Paris companion were lovers. I would be very surprised if this were true. Vittorio, who is a more lively character than the one portrayed in the film, was in fact a married man. Rosalind was friendly with both the Luzzatis, as she was later with Aaron Klug and his wife and with Odile and myself. I think Rosalind rather liked such friendships, since she could interact scientifically with the husband while enjoying the company of both partners. She would be friendly and relaxed without any danger of sexual involvement. Vittorio was at that time her closest scientific advisor, but he had little experience of solving structures of organic molecules in the Pauling manner so that his advice, though superficially sound, was in fact somewhat misleading.

The treatment has a number of interesting weaknesses. The scriptwriter, Bill Nicholson, was delighted to learn about the fiasco of our first model, because this seemed to fit a standard dramatic format. As he put it, "Boy meets girl; boy loses girl; boy gets girl," or, as he explained to me, a failure in the middle of the action gets the audience's sympathy on the side of the two "heroes." I could not help reflecting that when we made our blunder about the water content we were not trying to give dramatic shape to our efforts. We were hoping we had arrived at the correct structure.

The rapid cutting backward and forward between London and Cambridge as the climax approaches does correspond to the facts, even though the excitement is in a god's-eye view of the action, but the whole flavor of the ending has been distorted to make a theatrical climax. Although we were excited when we discovered the double helix, neither we nor anybody else thought of it as a wild success. Indeed Jim worried that it might be all wrong and that we'd again made fools of ourselves. Consequently the celebrations and the congratulations are figments of the scriptwriter's imagination. Most people would have described the structure as "interesting" or "very suggestive," but few would be confident at that stage that the double helix was really correct. Even less excusable is the "literary" twist given at the end. The idea that Jim was sobered (during the fictitious conversation on the bridge with his sister) because he had achieved all his aims is quite untrue to life. Moreover it fails to get over the true "ending"—that the double helix was not an ending but a beginning, because of all the ideas it suggested about gene replication, protein synthesis, and so on. This is what we worried about for the rest of the summer and for many years to come. Talk of prizes and success only came much later. When I returned to Cambridge from the States in the late summer of 1954 the Medical Research Council did not feel they had to give me tenure, although by then I was thirty-eight. They offered me a seven-year appointment, though a year or so later they converted this to an indefinite appointment (equivalent in the MRC to tenure).

As to the actors, I think Jeff Goldblum is too manic as Jim and far too interested in girls. "Nobody told me Jim didn't chew gum," Mick Jackson complained to me, but if he had looked carefully he would have discovered that almost no scientists chew gum, not even brash young American ones. Jim's natural manner was more subdued. Goldblum caught it rather well in the costume party scene, when he is asked if he is a real vicar (an Anglican clergyman). Incidentally, at the actual party Jim replied that he was. His questioner, a young American woman, quizzed him for half an hour about the spiritual upbringing of her children and was rather cross when she eventually discovered that he was not a clergyman at all.

As to the other actors, Max Perutz, Raymond Gosling, Maurice Wilkins, Peter Pauling, and Elizabeth Watson are immediately recognizable, but the really key performance is that of Juliet Stevenson as Rosalind. She is not only the true center of the film—she is almost the only person who really appears to be *doing* science—but we have a more complex inside view of her than of most of the other characters. I don't think this interpretation of Rosalind was an accident. Miss Stevenson's comments, quoted in the *Radio Times,* show that she had considerable insight into Rosalind's abilities and character. Moreover the scriptwriter has conveyed the general nature of Rosalind's error of judgment about the best method for solving the problem.

What, then, is one to make of *Life Story?* It certainly gets over the obvious fact that scientific research is performed by human beings, with all their virtues and weaknesses. There is no trace of the stereotyped emotionless scientist, solving problems by rigid logic. It shows, at least in outline, how one kind of science is done, though most research is more plodding and less dramatic than the discovery of the double helix. It even puts over, in an elementary way, a certain amount of basic scientific information. Most important of all, it tells a good story at a good pace, so that people from all walks of life can enjoy it and absorb some of these lessons in the process. All in all, in spite of its limitations, *Life Story* must be considered a success. In other hands it could easily have been nothing quite as good.

8

The Genetic Code

WITH THE DOUBLE HELIX clearly in view, the next problem was, what did it do—how did it influence the rest of the cell? We already knew the answer in outline. Genes determined the amino acid sequence of proteins. Because the backbone of the nucleic acid structure appeared so regular we assumed, correctly, that it was the base sequence that carried this information. Since the DNA was in the nucleus of the cell and since protein synthesis seemed to take place outside the nucleus, in the cytoplasm, we imagined that a copy of each active gene had to be sent to the cytoplasm. As there was plenty of RNA there, and no apparent trace of DNA, we assumed that this messenger was RNA. It was easy enough to see how a stretch of DNA would make an RNA copy—a simple base-pairing mechanism could do the trick—but it was less easy to see how the resulting messenger RNA (as we would now call it) could direct protein synthesis, especially as very little was then known about this latter process.

Moreover there was an informational problem. We knew there were about a couple of dozen different kinds of amino acids—the little units from which protein chains were made—yet there were only *four* different kinds of bases in DNA and RNA. One solution would be to read the nucleic acid sequence two bases at a time. This would yield only sixteen (4×4) possibilities, which seemed too few. Another alternative was to read them *three* at a time. This would give sixty-four ($4 \times 4 \times 4$) possible combinations of the four bases A, T, G, and C. This seemed too many.

It may help you to understand what follows if I outline our present knowledge of the genetic code. Unfortunately the phrase "genetic code" is now used in two quite distinct ways. Laymen often use it to mean the entire genetic message in an organism. Molecular biologists usually mean the little dictionary that shows how to relate the four-letter language of the nucleic acids to the twenty-letter language of the proteins, just as the Morse code relates the language of dots and dashes to the twenty-six letters of the alphabet.

I shall use the term in this latter sense. The details are set out in appendix B, which displays the little dictionary in the form of a table. The details of the table need not concern the lay reader. All you need to know is that the genetic message is read in non-overlapping groups of three bases at a time (for RNA, the bases being A, U, G, and C). Such a group is called a codon, a term invented by Sydney Brenner. It turns out that just twenty kinds of amino acids are coded for. In the standard code two amino acids have only one codon apiece, many have two, one has three, several have four, and two of them have six codons. In addition there are three codons for "end chain" ("start chain" is a bit more complicated). These add up to sixty-four codons in all. No codon is unused.

The proper technical term for such a translation rule is, strictly speaking, not a code but a cipher. In the same way the Morse code should really be called the Morse cipher. I did not know this at the time, which was fortunate because "genetic code" sounds a lot more intriguing than "genetic cipher."

An important point to notice is that although the genetic code has certain regularities—in several cases it is the first two bases that encode one amino acid, the nature of the third being irrelevant—its structure otherwise makes no obvious sense. It could well be that it is mainly the result of historical accidents in the distant past. Of course none of this was known in 1953 when the double helix was first discovered.

Jim and I had discussed the problem of protein synthesis in a desultory fashion that summer, but DNA itself was giving us so much to worry about—was the structure really correct? how exactly did it replicate itself?—that we had not seriously come to grips with it.

One day a letter arrived from America written in a large, round, unknown hand. We found we had already heard of its author, the physicist and cosmologist George Gamow, but the contents of the letter were quite new to us. Gamow had been intrigued by our papers in *Nature.* (Indeed we sometimes felt that physicists took more notice of them than biologists.) He jumped to the conclusion that the DNA structure itself was a template for protein synthesis. He noticed that, looked at in a certain way, the structure could have twenty different kinds of cavities, depending on the local sequence of the bases. Since there are about twenty different kinds of amino acids used to form the chains of proteins, he boldly assumed that there was just one type of cavity for each amino acid.

As we sat and studied Gamow's letter at the Eagle, Jim and I realized that we had never actually counted the exact number of types of amino acids found in proteins. It was not a completely straightforward matter, since there are many possible amino acids, only a few of them are found in living creatures, and not all of these occur in proteins. Protein chemists had discovered well over twenty amino acids in one protein or another, but some of these, such as hydroxyproline, were found in only one or two proteins and not in the general run of them.

Gamow had given his list of the magic twenty, but we immediately saw that some of them were unlikely and that he had left out some obvious candidates, such as asparagine and glutamine. There and then we wrote out our own list. I don't recall that Jim knew a lot about the finer points but fortunately I had by then acquired a detailed knowledge of many aspects of protein structure. The basic idea we used was that the amino acids that had been claimed to be in proteins should be classified either as members of a "standard" set or as "freaks." Any amino acid that was known to occur in many different proteins, such as alanine, was accepted as one of the standard set. An amino acid that occurred in only a few odd proteins, such as bromotyrosine, we classified as a freak. We also rejected any amino acid that, although it occurred in a polymer in the cell, had not yet been shown to exist in a true protein. Diaminopimelic acid, which is found in the cell walls of certain bacteria, fell into this class.

We did not insist that *every* protein had to have *all* the members of the standard set, since in a small protein one of the less common ones might be missing by chance, because its polypeptide chain contained rather few amino acids (the lack of tryptophan and methionine in insulin would be an example). To our astonishment, we arrived at exactly twenty. Rather remarkably, our list has turned out to be essentially correct. Unknown to us, Dick Synge, one of the inventors of modern chromatography, had drawn up a similar list, but his had one extra candidate—cystine as well as cysteine—which was fairly obviously unlikely.

It is worth noting that all the writers of biochemical textbooks had a much longer list. In the early part of the century, the discovery of a new amino acid that occurred in proteins was an important event. While those days were past, the glamour of the quest still hung around. A new amino acid, once its occurrence in a protein had been firmly established experimentally, was still deemed an important discovery and, as such, went into the textbooks. The idea that there might be a standard set of amino acids and that the rest were, in some sense, freaks had not penetrated to most biochemists, though obviously some protein chemists thought that way even if they had not formulated their ideas explicitly. We now know the proteins are synthesized by a very special mechanism that can handle only a limited number of amino acids. The others, the "freaks," are mostly standard amino acids that have been modified by extra processes after the polypeptide chain has been synthesized.

This is a nice example of complexity of nature produced by natural selection. It shows how easily one can be misled if one takes too straightforward a view of a biological problem. Of course, we were fortunate to have hit on the correct standard set at our first attempt. It was a lucky guess and needed to be confirmed by many additional experiments. While it took some years for biochemists to do this, that our list was correct was never seriously in doubt. Although there was occasional conflicting evidence, our list has stood the test of time. The only omission was the use of formylmethionine for chain initiation in prokaryotes, and this would have been impossible for us to foresee.

I cannot remember whether Gamow's first letter included a manuscript (I think this arrived a little later), but when we did get a copy—I still have it somewhere—we were surprised to see that Gamow had listed Tomkins as a coauthor. Gamow was well known as a popular science expositor, with a somewhat whimsical style. Mr. Tomkins, Gamow's Everyman, was a character in several of his books and usually appeared in the title (*Mr. Tomkins Explores the*

Atom, for example). Alas, before the paper was finally published the mythical Mr. Tomkins was removed by a stern editor.

Gamow's "code" was unusual in several ways. Each amino acid was coded by a triplet of bases (actually several triplets, related by symmetry), but the triplets standing for successive amino acids overlapped. For example, if a small part of the sequence was . . . GGAC . . . , then GGA stood for one amino acid, and GAC for the next one. Naturally this imposed restrictions on the amino acid sequence. Certain sequences could not be coded for by Gamow's code. The matter was not completely straightforward since Gamow did not know *which* of his triplets stood for *which* amino acid. This was left open, and would have had to be discovered by experiment. At that time, although the amino acid *composition* of many proteins had been determined, at least approximately, only fragments of *sequence* were known (Fred Sanger's complete sequence of the two chains of insulin were still in the works) so there was not much data with which to test Gamow's theory.

Jim and I had several objections to Gamow's ideas. We rather doubted whether the cavities in DNA were capable of doing the job. We worried about his symmetry assumptions, and we didn't like the idea of DNA coding directly for proteins. RNA seemed a more likely candidate, but perhaps RNA could fold up into a structure that could form the necessary cavities. Gamow had put in, implicitly, one restriction that seemed natural enough. When joined together in a chain, one amino acid is quite close to the next one—only about 3.7 Å apart (the distance between strongly bonded atoms is typically between 1 and 1 ½ Å). By contrast, a group of three bases spreads over a much larger distance. For this reason an overlapping code, which reduces this distance, seemed more likely, in spite of the restrictions it put on the possible amino acid sequences.

Gamow had made another contribution. We eventually realized that solving the code could be viewed as an abstract problem, divorced from the actual biochemical details. Perhaps by studying the restrictions on the amino acid sequences, as they became available, and by watching how mutants affected a particular sequence, one could crack the code without having to know all the intervening biochemical steps. Such an approach seems natural to a physicist, confronted by the complexities of chemistry and biochemistry, though in fairness to Gamow one must concede that his ideas were originally based on our model of the double helix, not just on abstract ideas.

That winter (1953–54), while I was working at the Brooklyn Polytechnic—it was my first visit to the States—I managed to disprove all possible versions of Gamow's code, by using the small amount of sequence data then available and by assuming (a quite unsupported assumption) that the code was "universal"—that is, was the same in all living organisms.

During the next summer Jim and I spent three weeks together at Wood's Hole. Gamow and his wife were there, staying at Albert Szent-Györgyi's cottage by the water. (Szent-Györgyi, a Hungarian, was awarded a Nobel Prize in 1937 mainly for discovering vitamin C.) By that time Gamow had come to know a number of people interested in the coding problem, in particular Martynas Ycas and Alex Rich. On most afternoons Jim and I went out to the cottage and sat on the shore with Gamow, discussing all the different aspects of the coding problem, idly chatting or just watching Gamow showing some of his card tricks to any pretty girl who happened to be around. The pace of scientific life in those days was less hectic than it is now.

By this time we knew Gamow well enough to call him Joe. His first name was George, but he signed his letters "Geo." He was under the impression that this was pronounced Joe, so that was what his friends called him. We were familiar with his boyish handwriting, his very Russian omission of articles *(a* and *the),* and his erratic spelling. We assumed that the latter was due to his writing in a foreign language, but later we learned that in his native Russian his spelling was just as bad. We were also impressed by his automobile, a large white convertible with red seats. He told me that a third of his income came from his academic salary, a third from writing, and a third from consulting, which partly explained his somewhat expensive car. He was fun to be with, and friendly, in spite of being older and more senior than we were. He was the champion of the Big Bang theory of the origin of the universe—among other things he predicted the existence of the background radiation, which had yet to be discovered. The Catholic Church preferred his theory to the rival theory of Continuous Creation, proposed by Gold, Bondi, and Hoyle. Even so, I was mildly surprised when he told me that he had exchanged reprints with the Pope, by way of the Holy Office.

Gamow enjoyed his glass of whiskey. Although I didn't realize it at the time, he was probably already on the slippery path to alcoholism. I was not at all surprised to receive by mail an invitation, in his own characteristic handwriting, to a "whiskey, twisty RNA Party" to be held at the cottage in a few days' time. The next time I went there I thanked Joe for his invitation, but he knew nothing about it. To his puzzlement letters of acceptance kept pouring in, brought down from the main house by Albert Szent-Györgyi. Naturally Joe suspected that Szent-Györgyi was the culprit, but he denied this. "On my heart," he said, "it is not me." Joe was embarrassed so I realized something had to be done. It did not take me long to discover that Jim was one of the perpetrators of the hoax. He did not usually play practical jokes, but his mentor, Max Delbrück, was notorious for them. The other hoaxer turned out to be Szent-Györgyi's nephew, Andrew Szent-Györgyi. I negotiated a treaty. Jim and Csuli, as he was known, would provide the beer and Joe would provide the whiskey. The party turned out to be a great success, with almost everyone invited turning up for it.

Meanwhile Joe, in his typical way, had founded that unusual organization, the RNA Tie Club. This was a very select club—Gamow decided who was to be a member. There were to be only twenty members, one for each amino acid, and not only did each member receive a tie, made to Gamow's design by a haberdasher in Los Angeles (Jim Watson and Leslie Orgel arranged this), but also a tie pin with the short form of his own amino acid on it. I think I was Tyr but I'm not sure I ever got the tie pin. The club never met, but it had notepaper that listed its officers. Geo Gamow was described as Synthesizer, Jim Watson as Optimist, and I as Pessimist. Martynas Ycas was denoted Archivist and Alex Rich as Lord Privy Seal. As it turned out the club served as a mechanism for circulating speculative manuscripts to the few people interested. After I returned to England in the fall of 1956 I wrote a paper for it analyzing Gamow's ideas, generalizing them, and suggesting what turned out to be an important idea, the adaptor hypothesis.

The paper was called "On Degenerate Templates and the Adaptor Hypothesis." The main idea was that it was very difficult to consider how DNA or RNA, in any conceivable form, could provide a direct template for the side-chains of the twenty standard amino acids. What any structure *was* likely to have was a specific pattern of atomic groups that could form hydrogen bonds. I therefore proposed a theory in which there were twenty adaptors (one for each amino acid), together with twenty special enzymes. Each enzyme would join one particular amino acid to its own special adaptor. This combination would then diffuse to the RNA template. An adaptor molecule could fit in only those places on the nucleic acid template where it could form the necessary hydrogen bonds to hold it in place. Sitting there, it would have carried its amino acid to just the right place it was needed.

There were several implications of this idea. The one I want to stress here was that it meant that the genetic code could have almost *any* structure, since its details would depend on which amino acid went with which adaptor. This had probably been decided very early in evolution and possibly by chance. Because of this pessimistic conclusion the paper led off with a quotation from an obscure Persian writer of the eleventh century: "Is there anyone so utterly lost as he that seeks a way where there is no way?" and ended with the remark, "In the comparative isolation of Cambridge, I must confess there are times when I have no stomach for the coding problem."

The paper was circulated to members of the RNA Tie Club but was never published in a proper journal. It is my most influential unpublished paper. Eventually I did publish a short remark briefly outlining the idea and tentatively suggesting that the adaptor might be a small piece of nucleic acid. It soon turned out that a biochemist at the Harvard Medical School, Mahlon Hoagland, had quite independently obtained some experimental evidence that supported my proposal. As every molecular biologist now knows, the job is done by a family of molecules now called transfer RNA. Ironically, I did not immediately recognize that these transfer RNA molecules were the predicted adaptor because they were considerably bigger than I had expected, but I soon saw that there were no grounds for my objection. A little later Mahlon came to Cambridge for a year and we did experiments together on transfer RNA. We worked in a small upstairs room in the Molteno Institute that the director graciously allowed us to use since it was temporarily vacant.

Much theoretical effort during this period was put into attempts to solve the coding problem, especially by Gamow, Ycas, and Rich. Gamow and Ycas suggested a "combination code" in which the *order* of the bases in a triplet did not matter, only its combination of bases. While this was structurally implausible it had some appeal because it so happens there are just twenty combinations of four things taken three at a time. Again there was no hint as to how to allocate each amino acid to its own combination.

For a time it was still thought that the code would have to be an overlapping one, and so the search for restrictions on the amino acid sequence continued. As new sequences became available they were added to those we had already collected, but there was little hint of any forbidden sequences, although the data were so sparse that at first we could not be sure that some sequences were missing. The hunt was mainly restricted to adjacent amino acids. There are 400 (20×20) possible amino acid doublets. *Any* overlapping triplet could code for only 256 (64 possible triplets $\times$ 4) of these, so there had to be restrictions if the code were of this type. Sydney Brenner realized that one could sharpen this argument. Any one triplet would have only four other triplets as its neighbors on one side. For example, if the triplet in question was AAT, then the only triplets that could precede it were *T*AA, *C*AA, *A*AA, and *G*AA, while only AT*T*, AT*C*, AT*A*, and AT*G* could follow it, assuming as always that the code was overlapping. Thus if in the known sequences one particular amino acid had been shown to have at least nine neighbors following it, then it would have to have at least three triplets allocated to it, since two triplets could have only eight neighbors following it. Sydney was able to show that the number of triplets needed easily exceeded sixty-four and thus that *all* overlapping triplet codes were impossible. This proof assumed that the code was "universal"—that is, was the same in all the organisms from which the experimental data had come—but this was sufficiently plausible to make us almost certain that the idea of an overlapping code was wrong.

This still left the geometrical dilemma. In the process of protein synthesis, how could one amino acid get near enough to the next one to enable them to be joined together, since their triplets would have to be some distance apart as they were not overlapping? Sydney suggested that the postulated adaptors might each have a small flexible tail, to the end of which the appropriate amino acid was joined. Sydney and I did not at the time take this idea very seriously, referring to it as a "don't worry" theory, meaning that we could see at least one way that nature might have solved the problem, so why worry at this stage what the correct answer actually was, especially as we had more important problems to tackle. In this case it has turned out that Sydney was correct. Each transfer RNA does indeed have a small flexible tail to which the amino acid is joined.

In parenthesis let me say that the English school of molecular biologists, when they needed a word for a new concept, usually use a common English word such as "nonsense" or "overlapping," whereas the Paris school like to coin one with classical roots, such as "capsomere" or "allosterie." Ex-physicists, such as Seymour Benzer, enjoyed inventing new words ending in "-on," such as "muton," "recon," and "cistron." These new words often obtained rapid currency. I was once persuaded by the molecular biologist François Jacob to give a talk to the physiology club in Paris. It was then the rule that all such talks had to be given in French. As I hardly speak French I did not warm to his suggestion at all, but François pointed out to Odile (who is bilingual in French and English) that if I gave the talk she also could have a trip to Paris, so my opposition was soon worn down. I decided to talk on the problem of the genetic code, thinking, quite incorrectly, that I could do most of it by simply writing on the blackboard. It soon became clear that I would have to speak some French in order to get the ideas across, so I started by dictating the whole talk to a secretary (normally I speak from notes). I then deleted all the jokes, since even when giving a talk to a secretary I found that my ad lib jokes intruded, and I felt I could hardly read them out in cold blood. Odile then translated the talk into French, and a typed version of her manuscript was produced, with various stress marks added to make it easier for me to read. There was a problem, however, about the translation of "overlapping." What could be the French for that? Odile eventually remembered a suitable word, and we set off for Paris. I was sufficiently mistrustful of this strange word that on arrival I asked François what word *they* used for "overlapping." "Oh," he said, "we simply say 'oh-ver-lap-pang.' "

I would like to report that the talk was a success. I started off fairly well, reading carefully, but as I warmed up my pronunciation got gradually wilder and wilder. The discussion, mainly in French, taxed me greatly. After the talk I asked François how it went. "It was not *too* bad," he said tactfully, "but it was not *you.*" With no spontaneity and no jokes I saw just what he meant. I have never since attempted to give a talk in a foreign language, even though my French accent has improved a little over the years.

It was now clear that the code was not overlapping, but this immediately raised a new problem. If the code was read as a sequence of *non*-overlapping triplets, how did we know where the triplets began? Put another way, if we were to imagine that the correct triplets were marked by commas (for example, ATC,CGA,TTC, . . .), how did the cell know exactly where to put the commas? The obvious idea, that one started at the beginning (whatever that was) and went along three at a time, seemed too simple, and I thought (quite wrongly) that there must be another solution. It occurred to me to try to construct a code with the following properties. If read in the right phase, all the triplets would be "sense" (that is, stand for one amino acid or another), whereas all the out-of-phase triplets (those that bridged the imaginary commas), would be "nonsense"—that is, there would be no adaptor for them and thus they would not stand for any amino acid. I mentioned this idea to Leslie Orgel, who immediately pointed out that for such a code the *maximum* number of sense triplets was twenty. A triplet such as AAA must be nonsense since otherwise the sequence AAA,AAA could be read out of phase. (We tacitly assumed by now that any amino acid could follow any other amino acid.) That eliminated four of the sixty-four triplets. If the XYZ triplet was sense, then the cyclic permutations YZX and ZXY would have to be nonsense, so the maximum number of sense triplets was 60/3 = 20. The problem was: Did a set of twenty triplets exist that had this property? I was confined to bed with a nasty cold but found I could easily get up to seventeen. Leslie mentioned the problem to John Griffith, who found a set of twenty with the right properties. We soon found several other solutions (plus numerous permutations) so there was no doubt that such a code could exist. We even invented a plausible argument why it could be useful.

The problem of finding a solution having twenty sense triplets is actually not an especially difficult one. A little later I was booked on a night flight from the States to England. Waiting to board I found myself chatting to Fred Hoyle, the cosmologist. He asked what I was doing and I explained to him the idea of the comma-free code. The next morning, as the plane approached the English coast, he came back to where I was sitting with a solution he had worked out overnight.

Naturally Orgel, Griffith, and I were excited by the idea of a comma-free code. It seemed so pretty, almost elegant. You fed in the magic numbers 4 (the 4 bases) and 3 (the triplet) and out came the magic number 20, the number of the amino acids. Without more ado we wrote it up for the RNA Tie Club. Nevertheless I was hesitant. I realized that we had no *other* evidence for the code, other than the striking emergence of the number twenty. But then if some other number had come up we would have discarded the idea and looked around for some other code that led to twenty amino acids, so the number twenty by itself was not confirmatory evidence.

In spite of my worries, the new code attracted some attention. After four people had asked if they could quote our paper (an RNA Tie Club note was not equivalent to publication), we decided to write it up for the *Proceedings of the U.S. National Academy of Science,* where it duly appeared in 1957. An account of it even appeared in a book for the general reader called *The Coil of Life* written by Ruth Moore, though this was not published till 1961, by which time we had ceased to believe in the idea.

Since in the comma-less code each amino acid had just one triplet it would have been possible, knowing which amino acid went with each triplet, to deduce the base composition of the DNA, assuming it all coded for protein, from the average amino acid composition of all its proteins. Because the latter was pretty similar in all organisms (though we knew now there were small variations), this would imply that the DNA molecules in all species had much the same composition. As more measurements were made, especially on different types of bacteria, it became clear that this was very far from the case. Of course in all cases the amount of A was the same as the amount of T (A=T) since the base pairing demanded this, and for the same reason G=C, but the structure of DNA itself put no restrictions on the ratio of A+T to G+C, and this ratio was found to vary a lot from one organism to another. This made it likely that the comma-free code must be wrong.

Its final downfall came from two directions. Our work on phase-shift mutants, described in chapter 12, made it unlikely, but a more decisive blow was dealt by Marshall Nirenberg when he showed that poly U (a simple form of RNA) coded for polyphenylalanine (see page 130), whereas in a comma-free code UUU should have been a nonsense triplet. Finally the correct genetic code, confirmed by so many methods, has proved decisively that the whole idea is quite erroneous. However, it is just conceivable that it may have played a role near the origin of life, when the code first began to evolve, but this is pure speculation.

The idea of comma-free codes attracted the attention of combinatorialists, in particular Sol Golomb. We had failed to solve the problem of enumerating all possible triplet overlapping codes (with four letters) although we had found more than one solution. This enumeration was worked out by Golomb and Welch, using a very neat argument (which we ought to have seen for ourselves) as a key part of the proof. The problem was also solved by the Dutch mathematician H. Freudenthal at about the same time.

Eventually the code (see appendix B) was solved by experimental methods, not by theory. Major contributors were the groups of Marshall Nirenberg and of Gobind Khorana. The group of an earlier Nobel laureate, Severo Ochoa, also made important contributions. Even as the code was coming out, attempts were made to guess the whole from the part, but these were also largely unsuccessful. In some ways the code embodies the core of molecular biology, just as the periodic table of the elements embodies the core of chemistry, but there is a profound difference. The periodic table is probably true everywhere in the universe, and especially relevant in places that have about the same temperature and pressure as the Earth. If there is life on other worlds and even if that life also uses nucleic acids and proteins, which is far from certain, it seems very probable that the code there would be substantially different. There are even minor variants of it in some of the organisms we have here on the Earth. The genetic code, like life itself, is not one aspect of the external nature of things but is, at least in part, the product of accident.

9

Fingerprinting Proteins

IN THE LAST CHAPTER I discussed the various theoretical attempts to solve the coding problem. In this one I describe some experimental approaches. The problem was much the same as before: Do genes (DNA) control the synthesis of protein? And if so, how?

It seems obvious enough now that the amino acid sequence of a protein is determined genetically, and in particular by the base sequence of a stretch of DNA (or RNA), but this was not always so clear. After the double helix was discovered the idea seemed much more attractive, so much so that Jim and I began to take it for granted. The next step was to show that the gene and the protein it coded were co-linear. By this I mean that the sequence of bases in that stretch of nucleic acid was in step with the corresponding sequences of amino acids in the particular protein it coded, just as a stretch of Morse code is co-linear with the corresponding message in English.

In those days there seemed no hope of sequencing either DNA or RNA directly, but in favorable circumstances we thought it might be possible to order a set of mutants within one gene, using standard genetic methods. Since the genetic distances were likely to be rather small, the recombination rates involved were expected to be much less than those geneticists usually measured. This implied that many progeny would have to be examined, suggesting that it would be necessary to use some sort of microorganism, such as a bacterium or a virus.

Once the mutants had been put in order, the next step would be to pin down the amino acid change due to each mutant. Although sequencing a protein chain was then still laborious, Fred Sanger had shown that it could be done, and we expected that for a small protein it would not be impossibly difficult.

Some time in the summer of 1954 I was sitting on the grass at Wood's Hole, explaining these ideas to the Polish geneticist Boris Ephrussi. Boris, by then working in Paris, had been particularly interested in genes in yeast that appeared to be outside the nucleus of the cell. We know now that such cytoplasmic genes are coded in the DNA of the cell's mitochondria, but at that time all that was known was that they did not behave like nuclear genes. Boris was indignant. "How do you know," he asked, "that the amino acid sequence is not determined by a cytoplasmic gene and that all the nuclear genes do is to fold up the protein correctly?"

I don't think Boris necessarily believed this (and certainly I did not), but his question made me realize that we first needed to show that a *single* mutant in a nuclear gene altered the amino acid sequence of the protein for which it coded, probably changing just a single amino acid. On returning to Cambridge I decided that this was the next most important step to take.

It was not at all clear what organism to use nor what protein to study. A little later Vernon Ingram joined us at the Cavendish. His main task was to add heavy atoms to hemoglobin or myoglobin, to help the X-ray work, but he and I decided to have a go at the genetic problem. We realized that for the first step we need not map the gene in detail. All we needed was enough genetic information to show that a mutant was being inherited in a Mendelian way and was therefore likely to belong to a nuclear gene. Nor did we need to fix the changed amino acid in the sequence. It was only necessary to show that there had been a change in the sequence due to the mutant. We thought that this would make things easier, since we then only needed to study the amino acid *composition* of the proteins. If the protein were small enough we might, with luck, pick up a change as small as an alteration to just one amino acid.

In order to work with a protein that was easy to obtain, we chose the protein lysozyme. Lysozyme is a small, basic (meaning positively charged) enzyme originally characterized by Alexander Fleming, the discoverer of penicillin. Fleming had shown that it occurred in tears and that egg white was also a rich source. The enzyme lyses (breaks up) a certain class of bacteria, and in both contexts acts to counteract bacterial infection. One particular bacterium is especially sensitive to it, and this can be used as an assay for the enzyme.

Our main target was egg white but we also tried human tears. Each morning when I came into the laboratory the assistant took a small sample of my tears. Not being an actor, I did not find it easy to weep at will, so my assistant would hold a slice of raw onion underneath one eye. I would hold my head to one side, to make it less easy for the tear to escape down the tear duct, and she would catch the tears with a little Pasteur pipette as they dribbled out of the other side of my eye. Even so, it was difficult to produce more than one or two tears, though I found it helped to think sad thoughts. Curiously enough, I never cry spontaneously at sad or tragic events, but a happy ending makes me weep uncontrollably. Let the bride finally walk triumphantly down the aisle, with the organ playing in jubilation. The tears will stream down my face, in spite of my intense annoyance and embarrassment.

The effect of a single tear can be dramatic. A weak suspension of the bacteria we used looks appreciably cloudy, though not as dense as milk. Add a single tear, swirl the fluid in the test tube, and in a moment the suspension becomes completely clear. All the bacteria have been lysed, thus immediately reducing the scattering of light that caused the cloudiness. Of course we used a more quantitative assay, but the phenomenon was basically the same.

Because chick lysozyme has a strong positive charge, unlike all the other proteins in egg white, it is possible to crystallize it *in the egg white,* without any further purification. To a biochemist it is really surprising to see the crystals sitting in the rather concentrated, gooey egg white. For the same reason lysozyme was relatively easy to separate on the simple ion exchange columns that had just then been developed for fractionating proteins.

It would be nice to report that we found a mutant, but in fact we had no success at all. We tested the lysozyme rather crudely, checking, in effect, its charge and the way it absorbed ultraviolet light, yet we could easily show that chick lysozyme differed from guinea fowl lysozyme, and that they were both quite different from the lysozyme in my tears. Although we studied about a dozen strains of chickens, kindly supplied by the local chicken geneticist, testing about a hundred eggs in all, we never detected any difference. We tried the tears of half a dozen people around the lab, but these all seemed to be similar to each other. I wanted to test the tears of my younger daughter Jacqueline, then only two years old, but Odile would have none of it. What! Use her precious baby for an experiment! I was sternly forbidden to attempt it.

I expect we would have gone on, but at that stage there was a dramatic development. Max Perutz was working on hemoglobins, including human hemoglobin. Some years earlier Harvey Itano and Linus Pauling had shown that the hemoglobin from a person with sickle-cell anemia was electrophoretically different from normal hemoglobin. Pauling rightly dubbed it a genetic disease. A colleague of his at Cal Tech measured its amino acid composition and reported that there was no difference between normal and sickle-cell hemoglobin. This conclusion was badly worded. What he meant was that there was no difference in composition he could reliably detect, but since hemoglobin is a comparatively large protein, a single amino acid change could easily be missed using this rather crude measure.

Sanger had developed a method he called fingerprinting proteins. He digested the protein with an enzyme (trypsin) that cut the polypeptide chain only at special places. The limited number of peptide fragments thus produced were then run on a two-dimensional paper chromatographic system to sort them one from another, spreading the peptides out on the paper. Vernon realized that this was just the method he needed to pick up small alterations in a protein. Fortunately Max had been sent some sickle-cell hemoglobin, and he gave some to Vernon to test. To his delight, the fingerprints of sickle-cell hemoglobin and of normal hemoglobin differed in the position of a single peptide.

Vernon was able to isolate the altered peptide, determine its sequence, and show that indeed the difference was due to the change of a single amino acid. Valine had been substituted for glutamic acid. At one point, I recall, he thought that perhaps two amino acids might be changed. Jim and I were brasher then and refused to believe this. "Try it again, Vernon," we said, "you'll find there's just a single change" and so it turned out to be.

This result was surprising from two points of view. Sickle-cell anemia is a disease in which the altered hemoglobin forms a type of crystal inside the "red" cells of the blood when it gives up its oxygen in the veins. This often breaks the red cell open, so that patients have a chronic lack of hemoglobin in their blood and, in many cases, die in their teens. Yet this lethal effect is produced by a tiny alteration in just one of the organism's many genes (we know now it is due to a single base change). Essentially just two molecules are defective, one inherited from the father and one from the mother. How can such a minute change possibly kill someone? The reason is the cascade of magnification. Each defective gene is copied many, many times, since each cell in the body has to have its own copy. Then, in the precursors of each red cell, each gene is copied many times onto messenger RNA, and each messenger RNA directs the synthesis of many defective protein molecules. The tiny atomic defect gets magnified and magnified till there is a considerable amount of the defective protein in the patient's body, quite enough to kill him if the circumstances are unfavorable.

The other surprising aspect was the scientific one. Strange as it may seem, up to that point most geneticists and protein chemists had not seriously considered that their respective fields were related. Of course a few farsighted individuals, such as Hermann Muller and J. B. S. Haldane, were aware of the likely connection, but each field pursued its aims with very little awareness of the other. Ingram's result produced a dramatic change of attitude. At about this time I ran into Fred Sanger, I think on a train to London. He said that he and his small group thought they ought to learn a little genetics, a subject about which, up to that point, they hardly knew anything at all except that it existed.

I arranged that we should have weekly evening meetings in my sitting room at the Golden Helix. Sydney Brenner and Seymour Benzer agreed to conduct these tutorials. I recall the first one rather vividly. Sydney came over a little while before the others. I asked him what he proposed to say. He said he thought he would start with Mendel and peas. I suggested that this was perhaps by now a little old-fashioned. Why not start with haploid organisms (which have only one copy of the genetic material), such as bacteria, rather than peas or mice or men, which are diploid (that is, with two copies in each cell) and thus more complicated? Sydney agreed. He gave a brilliant lecture, mainly on the difference between genotype and phenotype, illustrated with examples from bacteria and bacterial viruses. It was all the more striking since I knew it was improvised as he went along.

I think that there is a lesson here for those wanting to build a bridge between two distinct but obviously related fields (a possible modern example would be cognitive science and neurobiology). I am not sure that reasoned arguments, however well constructed, do much good. They may produce an awareness of a possible connection, but not much more. Most geneticists could not have been easily persuaded to learn protein chemistry, for example, just because a few clever people thought that was where genetics ought to go. They thought (as functionalists do today) that the logic of their subject did not depend on knowing all the biochemical details. The geneticist R. A. Fisher once told me that what we had to explain was why genes were arranged like beads on a string. I don't think it ever occurred to him that the genes made up the string!

What makes people really appreciate the connection between two fields is some new and striking result that obviously connects them in a dramatic way. One good example is worth a ton of theoretical arguments. Given that, the bridge between the two fields is soon crowded with research workers eager to join in the new approach.

10

Theory in Molecular Biology

As WE HAVE JUST SEEN, the genetic code was a problem that would not yield to purely theoretical approaches. This does not mean that some general theoretical framework could not be helpful, if only to guide the directions that experiments might take. It was the nature of the structure of DNA that gave life to such speculations. Otherwise they would have been too vague to be useful. In 1957 I was invited to give a paper to a symposium of the Society for Experimental Biology in London. This gave me the opportunity to sort out and write down my ideas, most of which had been formulated earlier.

What the structure of DNA suggested was that the sequence of bases in the DNA coded for the sequence of amino acids in the corresponding protein. In the paper I called this the sequence hypothesis. Rereading it, I see that I did not express myself very precisely, since I said ". . . it assumes that the specificity of a piece of nucleic acid is expressed solely by the sequence of its bases, and that this sequence is a (simple) code for the amino acid sequence of a particular protein." This rather implies that *all* nucleic acid sequences must code for protein, which is certainly not what I meant. I should have said that the only way for a gene to code for an amino acid sequence of a protein is by means of its base sequence. This leaves open the possibility that parts of the base sequence can be used for other purposes, such as control mechanisms (to determine if that particular gene should be working and at what rate) or for producing RNA for purposes other than coding. However, I don't believe anyone noticed my slip, so little harm was done.

The other theoretical idea I proposed was of a rather different character. I suggested that "once 'information' has passed into protein *it cannot get out again,*" adding that "Information means here the precise determination of sequence, either of bases in the nucleic acid or of amino acid residues in the protein" (see appendix A).

I called this idea the central dogma, for two reasons, I suspect. I had already used the obvious word hypothesis in the sequence hypothesis, and in addition I wanted to suggest that this new assumption was more central and more powerful. I did remark that their speculative nature was emphasized by their names.

As it turned out, the use of the word dogma caused almost more trouble than it was worth. Many years later Jacques Monod pointed out to me that I did not appear to understand the correct use of the word dogma, which is a belief *that cannot be doubted.* I did apprehend this in a vague sort of way but since I thought that *all* religious beliefs were without any serious foundation, I used the word in the way I myself thought about it, not as most of the rest of the world does, and simply applied it to a grand hypothesis that, however plausible, had little direct experimental support.

What is the use of such general ideas? Obviously they are speculative and so may turn out to be wrong. Nevertheless, they help to organize more positive and explicit hypotheses. If well formulated, they can act as a guide through a tangled jumble of theories. Without such a guide, any theory seems possible. With it, many hypotheses fall away and one sees more clearly which ones to concentrate on. If such an approach still leaves one lost in the jungle, one tries again with a new dogma, to see if that fares any better. Fortunately in molecular biology the one first selected turned out to be correct.

I believe this is one of the most useful functions a theorist can perform in biology. In almost all cases it is virtually impossible for a theorist, by thought alone, to arrive at the correct solution to a set of biological problems. Because they have evolved by natural selection, the mechanisms involved are usually too accidental and too intricate. The best a theorist can hope to do is to point an experimentalist in the right direction, and this is often best done by suggesting what directions to avoid. If one has little hope of arriving, unaided, at the correct theory, then it is more useful to suggest which class of theories are *un*likely to be true, using some general argument about what is known of the nature of the system.

Looking back, it can now be seen that "On Protein Synthesis" is a mixture of good and bad ideas, of insights and nonsense. Those insights that have proved correct are the ones based mainly on general arguments, using data established for some time. The incorrect ideas sprang mainly from the more recent experimental results, which in most cases have turned out to be either incomplete or misleading, if not completely wrong.

Even at this stage an erroneous idea had crept in. It is clear that I thought of the RNA in the cytoplasm—in the microsomal particles, as they were then called (the word ribosome had not yet come into general use)—as a "template"; that is, as having a rather rigid structure, comparable to the double helix of DNA though probably having only a single chain. It was only later that I realized that this was too restrictive an idea, and that "tape" might be nearer the truth. Just as a ticker tape has no rigid structure except momentarily when it is actually in the ticker machine, I eventually realized that the RNA directing the synthesis of a protein need not be rigid, but could be flexible, except for that part that coded the next amino acid to be incorporated. Another consequence of this idea was that the growing protein chain did not have to stay on the template but could start to fold itself up as synthesis proceeded, as indeed had been suggested earlier.

There was another more serious mistake in my thinking at that time. I will not spell out all the details (they are given in the paper), but in effect I was making mistakes because I was confusing the mechanism itself (of protein synthesis) with completely separate mechanisms that were controlling it. Thus, in brief, because some experiments suggested that free leucine (one of the amino acids) was needed for RNA synthesis, it was concluded that there were probably common intermediates for protein and RNA synthesis, which could be used to synthesize one or the other, as required. In fact it is the *control* mechanism that requires free leucine if RNA synthesis is to continue, presumably because new RNA is not needed if the cell is so starved that free leucine is not available. I believe one can easily fall into this mistake of mixing up effects due to the nature of a mechanism itself with effects due to its control when trying to unscramble a complex biological system.

Another mistake in this general category is worth noting at this point. This is to mistake a minor process, evolved to improve the performance of the major process, for the major process itself and hence draw false conclusions about the latter. Alternatively one can be ignorant of the minor process and hence conclude that a postulated mechanism for the major process could not work.

Consider, for example, the rate of making errors in DNA replication. It is not difficult to see that if an organism has a million significant base pairs, then the error rate per step of replication should not be as great as one in a million. (The exact formulation has been given rather elegantly by Manfred Eigen.) Human DNA has about three billion base pairs (per haploid set) and although we now know that only a fraction of these have to be replicated accurately, the error rate cannot be greater than about one in a hundred million (speaking very roughly) or the organism would be torpedoed in evolution by its own errors. Yet there is a natural rate for making replication errors [due to the tautomeric nature of the bases] that it would be difficult to reduce to below about one in ten thousand. Surely, then, DNA cannot be the genetic material since its replication would produce too many errors.

Fortunately we never took this argument seriously. The obvious way out is to assume that the cell has evolved error-correcting mechanisms. Because the double helix carries two (complementary) copies of the sequence information, it is easy to see how this might be done. The *observed* error rate (the mutation rate) would be due to the errors in the error-correcting mechanism and thus can be reduced to a very low value. Leslie Orgel and I actually wrote a private letter to Arthur Kornberg, pointing this out and predicting that the enzyme he was studying that replicated DNA in the test tube (the so-called Kornberg enzyme) should contain within itself an error-correcting device, as indeed it does. DNA is, in fact, so precious and so fragile that we now know that the cell has evolved a whole variety of repair mechanisms to protect its DNA from assaults by radiation, chemicals, and other hazards. This is exactly the sort of thing that the process of evolution by natural selection would lead us to expect.

There is perhaps one other type of mistake that is worth mentioning. One should not be too clever. Or, more precisely, it is important not to believe too strongly in one's own arguments. This particularly applies to *negative* arguments, arguments that suggest that a particular approach should certainly not be tried since it is bound to fail.

Consider the following example. As far as I know this argument was never made but it could easily have been in, say, 1950. Rosalind Franklin had shown that fibers of DNA, especially when pulled carefully and mounted under conditions in which the humidity was controlled, could give an X-ray diffraction pattern of the so-called A form, which has many fairly sharp spots. Using the theory of Fourier Transforms, it can be seen immediately that these spots show the existence of a structure with a regular repeat. If DNA were the genetic material it could hardly have a regular repeat, since it could carry no information. Thus DNA cannot be the genetic material.

However, there is a counterargument to this. The X-ray spots do not extend to very small spacings. Why do the spots fall off in this way? It could either be that the structure is highly regular but is distorted in some random manner in the fiber, or it could be that part of the structure is regular and part is irregular. If so, why should not the irregular part carry the genetic information? If this is the case, then solving the regular part of the X-ray structure, using the spots that do exist, will never tell us what we want to know—the nature of the *genetic* information—so why bother to do it?

Knowing the answer, the fallacy in this negative argument can be seen. It is true indeed that the X-ray data on fibers can never tell us the intimate details of the base sequence. What the data did lead to was the model of the double helix with base pairing as its key feature. At the low resolution associated with these spots, one base pair looks rather like any of the other three, but what the model showed us, for the first time, was the *existence* of base pairs, and this turned out to be crucial for the rapid development of the subject.

What, then, was the proper argument that should have been used? Surely it is that the chemical nature of genes is a subject of overwhelming importance. Genes were known to occur on chromosomes, and that was where DNA is found. Thus *anything* to do with DNA should be pursued as far as it can be, since one can never be sure in advance what may turn up. While one should

My father, Harry Crick, as a young man. (Author's collection.)

My mother, Anne Elizabeth Crick, in 1938. (Author's collection.)

Myself with my younger brother Tony. (Author's collection.)

My uncle Arthur Crick, who helped me financially. (Author's collection.)

Odile during the Second World War just before we met (Author's collection.

My son Michael, at Stockholm, 1962. (Author's collection.)

Our house, "The Golden Helix,"
19–20 Portugal Place, Cambridge.
(Author's collection.)

Myself with our two daughters,
Gabrielle (*left*) and Jacqueline,
taken about 1956.
(Author's collection.)

Francis and Odile Crick
invite you to
A SOUTH SEAS PARTY
at
The Golden Helix
on
Saturday, February 20th at 9 p.m.

DRESS

Natives
Planters
Missionaries ♂

♀ *Natives*
*Dancing Girls**
Goddesses

R.S.V.P.—19 Portugal Place

**Tights may be worn*
Planters' wives not admitted

Invitation to one of our parties, 1960.
(Author's collection.)

Jim Watson (*left*) with me in front of our demonstration model of the DNA double helix, summer 1953. (From J. D. Watson's *The Double Helix,* Atheneum, New York, 1968.)

Jim Watson, as he appeared in the August 1954 edition of *Vogue* "with the bemused look of a British poet." (Courtesy of Diana Edkin.)

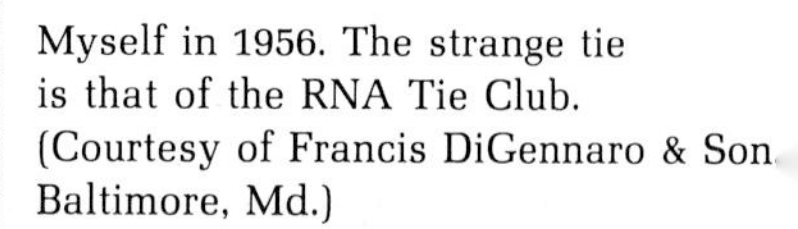

Myself in 1956. The strange tie is that of the RNA Tie Club. (Courtesy of Francis DiGennaro & Son, Baltimore, Md.)

A studio portrait of Rosalind Franklin, taken when she was about twenty-six. (Courtesy of Jenifer Glynn, Cambridge.)

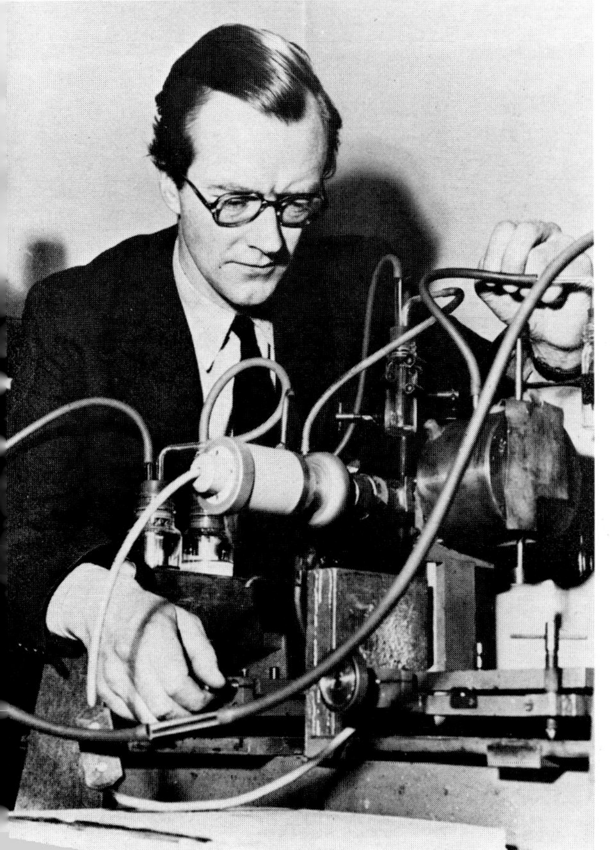

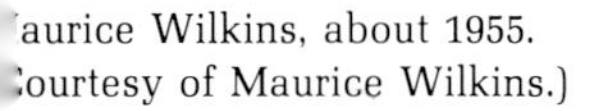

aurice Wilkins, about 1955.
ourtesy of Maurice Wilkins.)

J. D. Bernal, known to his friends as "Sage." (Courtesy of the Royal Society, London.)

Sir Lawrence Bragg—
Willie to his close friends.
(Courtesy of the Royal Society,
London.)

Linus Pauling in the 1950s,
holding models of two molecules.
(Courtesy of the Archives,
California Institute of Technology.)

Max Delbrück in conversation, June 1959.
(Courtesy of the Archives,
California Institute of Technology.)

Meeting of the RNA Tie Club
(*from left to right:* myself, Alex Rich,
Leslie Orgel, Jim Watson).
(Courtesy of Alex Rich, Cambridge, Mass.)

Max Perutz (*right*) handing over
the MRC Laboratory of Molecular Biology
to Sydney Brenner in 1979.
(Author's collection.)

Maurice Wilkins, Max Perut[z,]
myself, John Steinbec[k,]
Jim Watson, and John Kendre[w]
at the Nobel ceremony, 196[2.]
(Courtesy of Svensk[t]
Pressfoto, Stockholm[.)]

King Gustaf VI Adolf of Sweden,
with Odile at the Nobel banquet, 1962.
(Courtesy of Svenskt Pressfoto, Stockholm.)

Myself dancing with my elder daughter Gabrielle
at the Nobel Prize celebration, 1962.
(Courtesy of International Magazine
Service, Stockholm.)

certainly try to think which lines are worth pursuing and which are not, it is wise to be very cautious about one's own arguments, especially when the subject is an important one, since then the cost of missing a useful approach is high.

The example just given about DNA was a hypothetical one, but I have been caught in this way more than once. Experiments had shown that transfer RNA (tRNA) molecules existed, that amino acids were associated with them, and that there were probably many types of tRNA molecules, each with its own particular amino acid. The obvious next step was to purify at least one type of tRNA away from all the others so that more could be learned about it, as it was obviously better to work where possible on a pure species than on a mixture.

The problem was how to fractionate such a mixture. I argued to myself that since all tRNA molecules had to do a similar job and in particular to fit into the same place, or set of places, at the ribosome, they would all be very similar to each other and thus difficult to separate. The only way to separate them, I felt, was to use some method that tried to latch onto the amino acid joined to the RNA, by going for the particular side group of that amino acid and choosing one, such as cysteine, that was chemically both active and unique. I even tried to do this experimentally.

The argument was not totally silly, but it turned out I was wrong. Though I could not know it at the time, most tRNA molecules have many modified bases. These modifications alter their chromatographic behavior and so make it possible to separate them by much simpler fractionation methods since, in the first instance, only one of them is wanted. There is no need to specify in advance *which* tRNA to study, one simply experiments on the one that is easiest to get hold of. As the molecular biologist Bob Holley found, this turned out to be the tRNA for alanine, since it ran differently on a chromatography column from all the others. Again the message to experimentalists is: Be sensible but don't be impressed too much by negative arguments. If at all possible, try it and see what turns up. Theorists almost always dislike this sort of approach.

The path to success in theoretical biology is thus fraught with hazards. It is all too easy to make some plausible simplifying assumptions, do some elaborate mathematics that appear to give a rough fit with at least some experimental data, and think one has achieved something. The chance of such an approach doing anything useful, apart from soothing the theorist's ego, is rather small, and especially so in biology. Moreover I have found, to my surprise, that most theorists do not appreciate the difference between a model and a demonstration, often mistaking the latter for the former.

In my terminology, a "demonstration" is a "don't worry" theory (see the one described on page 97). That is, it does not pretend to approximate to the right answer, but it shows that at least a theory of that general type can be constructed. In a sense it is only an existence proof. Curiously enough, there exists in the literature an example of such a demonstration in relation to genes and DNA.

Lionel Penrose, who died in 1972, was a distinguished geneticist who in his later years held the prestigious Galton chair at University College, London. He was interested in the possible structure of the gene (which not all geneticists were at that time). He also loved doing "fretwork" (as it is called in England), making objects out of plywood with a fine saw. He constructed a number of such models to demonstrate how genes might replicate. The wooden parts had ingenious shapes, with hooks and other devices, so that when shaken they would come apart and join together in an amusing way. He published a scientific paper describing them and also a more popular article in *Scientific American.* An account by his son, Roger Penrose, the distinguished theoretical physicist and mathematician, appears in his father's obituary written for The Royal Society.

I was taken to meet Lionel Penrose and his models by the zoologist Murdoch Mitchison. I tried to show a polite interest but had some difficulty in taking it all seriously. What to me was bizarre was that this was in the middle 1950s, *after* the publication of the DNA double helix. I tried to bring our model to Penrose's attention but he was far more interested in his own "models." He thought that perhaps they might be relevant for a pre-DNA period in the origin of life.

His wooden pieces, as far as I could see, had no obvious relation to known (or unknown) chemical compounds. I cannot believe that he thought genes were made of pieces of wood, yet he didn't seem at all interested in organic chemicals as such. Why, then, was his approach of so little use? The reason is that his model did not approximate the real thing closely enough. Of course, any model is necessarily a simplification of some sort. Our DNA model was made of metal, but it embodied very closely the known distances between chemical atoms and, in the hydrogen bonds, took into account the different strengths of the various chemical bonds. The model did not itself obey the laws of quantum mechanics, but it embodied them to some extent. It did not vibrate, due to thermal motion, but we could make allowance for such vibrations. The crucial difference between our model and Penrose's was that ours led to detailed predictions on matters that had not been explicitly put into the model. There is perhaps no precise dividing line between a demonstration and a model, but in this case the difference is very clear. The double helix, since it embodied detailed chemical features, was a true model, whereas Penrose's was no better than a demonstration, a "don't worry" theory.

It was all the more odd that his "model" came well after ours. What was its fascination for him? I think, at bottom, he *liked* to do fretwork, to play with little pieces of wood, and he was delighted that his favorite hobby could be used to illuminate one of the key problems in his professional life—the nature of the gene. I suspect that, on the other hand, he disliked chemistry and didn't want to be bothered with it.

I cannot help thinking that so many of the "models" of the brain that are inflicted on us are mainly produced because their authors love playing with computers and writing computer programs and are simply carried away when a program produces a pretty result. They hardly seem to care whether the brain actually uses the devices incorporated in their "model."

A good model in biology, then, not only should address the problem in hand but if at all possible should serve to unite evidence from several different approaches so that various sorts of tests can be made of it. This may not always be possible to do straight away—the theory of natural selection could not immediately be tested at the cellular and the molecular level—but a theory will always command more attention if it is supported by unexpected evidence, particularly evidence of a different *kind.*

11

The Missing Messenger

THE NEXT EPISODE I want to touch on concerns what we now call messenger RNA. The double-helical structure of DNA had given us a theoretical framework that was invaluable as a guide to research, since it not only tied together approaches that at first sight seemed to have no connection with each other, but it suggested radically new experiments that could not have been conceived without the DNA model as a guide. Unfortunately, our thinking contained one major error. It was uncertain at that time whether any protein synthesis took place in the nucleus of the cell (where most of the DNA was), but everything suggested that the majority of it took place in the cytoplasm. In some way the sequence information in the nuclear DNA had to be made available outside the nucleus, in the cytoplasm. The obvious idea, which predated the DNA model, was that this messenger was RNA. This was the basis of the slogan coined by Jim Watson: "DNA makes RNA makes protein."

It was known that cells very active in protein synthesis had more RNA in their cytoplasm than cells that were less active. By the late 1950s it had been shown that most of their RNA was in small particles, now named ribosomes, that consisted of RNA molecules plus a mixture of proteins. What more natural than to assume that each ribosome synthesized just one protein and that its RNA was the postulated messenger RNA? We assumed that each active gene produced a (single-stranded) RNA copy of itself, that this was packaged in the nucleus with a set of proteins to help it do its job and then exported to the cytoplasm where it directed the synthesis of the particular polypeptide chain coded for by this RNA. Each ribosome, working in concert with the transfer-RNA molecules (see appendix A), would in some way embody the details of the genetic code (surmised, but not yet discovered) so that the four-letter language of the RNA could be translated into the twenty-letter language of the proteins.

About this time, Sydney Brenner and I discussed at some length how we could prove this idea by isolating a single ribosome, supplying it with all the necessary precursors, and then showing that it produced just one type of protein. Fortunately the problem seemed hopelessly difficult, as the techniques then available were not sensitive enough. We might have wasted much time and effort in difficult experiments that, unbeknown to us, were bound to fail.

Since ribosomes were obviously important structures, there was much experimental work on them. The techniques used were often new ones and thus open to suspicion, and the results were seldom clear-cut. Nevertheless, a whole series of awkward "facts" started to press for attention. The ribosomal RNA in a growing bacterial cell hardly seemed to turn over at all and was therefore described as "an inert metabolic product." RNA molecules in the ribosomes would have been expected to vary in length, since one protein is often a very different length from another. Yet the experiments indicated that the ribosomal RNA came in just two fixed sizes. The base composition of DNA in different species of bacteria varied over a wide range. Their messenger RNA might be expected to vary in the same way, yet the composition of the postulated messenger, the ribosomal RNA, varied only a little in these very different species. We could invent ad hoc reasons to explain away all these weak facts, but they made us very uncomfortable. Sydney and I spent many long hours going over the evidence, trying to spot what was wrong.

As it turned out, clarification came from a quite different source. The group of workers at the Institut Pasteur in Paris had carried out an experiment known as the PaJaMo experiment, because the authors were Arthur Pardee (a visiting American), Jacob, and Monod.

Monod's interest was mainly in the formation of induced enzymes and in particular in the enzyme β-galactosidase. The cell switched on the synthesis of this enzyme if the sugar galactose was supplied to it instead of the more customary glucose. Jacob's main concern was how genetic information was passed between cells during mating. He and Eli Wollman had done the famous blender experiment on bacteria in which the "male" and "female" cells had been allowed to join together and then, after a chosen time, had been separated by putting them in a Waring blender, an example of molecular coitus interruptus. Fortunately the process of mating is a prolonged one (it can last up to two hours, equivalent to several normal lifetimes for a rapidly growing cell), which makes it easier to study. They had shown that the genes were transferred in a linear fashion over this period, in a fixed order, so that interrupting the process had little effect on the earlier genes but prevented the transfer of the later ones. This turned out to be the key discovery in bacterial genetics, clearing up a whole series of complications and difficulties that had accumulated over the years.

From our point of view the significant aspect of this process was that a particular gene, such as the gene for β-galactosidase, would be introduced into the cell at a known time. It was then possible to see how the synthesis of this new protein changed with time, after the gene was introduced into the cell.

The result was surprising. We would have expected that the new gene started fairly soon to produce its own ribosomes, that these would accumulate slowly, and that as more and more ribosomes came into operation protein synthesis would steadily accelerate. The PaJaMo experiment showed something quite different. Very shortly after the gene was introduced the synthesis of β-galactosidase started up at a fairly fast rate and stayed that way.

Naturally we were reluctant to believe this experiment. Jacques Monod had first told us about it when he visited Cambridge, but at that stage the results were preliminary. Sydney and I worried about it during the following months. I tried to devise some way out, but my attempts seemed very forced.

A little later François Jacob came to Cambridge and on Good Friday, 1960, when the laboratory was closed, a small group of us assembled in a room in the Riggs Building of King's College, of which Sydney was a Fellow. Horace Judson has given a much fuller account of the whole episode. Here I will only touch on the main points.

The Missing Messenger

I started by cross-examining François about the PaJaMo experiment, since there were several possible loopholes in the original paper. François detailed to us how the experiments had been improved. He also reported a very recent experiment by Pardee and Monica Riley in Berkeley. Slowly we realized that we would have to accept the results as correct. Exactly what happened then is obscure, since it has been obliterated in what follows, but the train of thought can easily be reconstructed. What the PaJaMo type of experiment showed was that the ribosomal RNA could not be the message. All the previous difficulties had prepared us for this idea, but we had not been able to take the necessary next step, which was: Where, then, is the message? At this point Sydney Brenner let out a loud yelp—he had seen the answer. (So had I, for that matter, though nobody else had.) One of the peripheral problems of this confused subject had been a minor species of RNA that occurred in *E. coli* shortly after it had been infected by bacteriophage T4. (*E. coli,* a bacterium that lives in our gut, is much used in the laboratory.) Some years earlier, in 1956, two workers, Elliot Volkin and Lazarus Astrachan, had shown that a new species of RNA was synthesized that had an unusual base composition, since it mirrored the base composition of infecting phage and not that of the *E. coli* host, which happened to be very different. They had at first thought that this might be the precursor of the phage DNA, which the infected cell was compelled to synthesize in large quantities, but further careful work on their part had shown that this hypothesis was incorrect. Their result had hung in midair, surprising but unexplained.

The problem was then: If messenger RNA was a different species of RNA from ribosomal RNA, then why had we not seen it? What Sydney had seen was that the Volkin-Astrachan RNA *was* the messenger RNA for the phage-infected cell. Once this key insight had been obtained, the rest followed almost automatically. If there was a separate messenger RNA, then clearly a ribosome did not need to contain the sequence information. It was just an inert reading head. Instead of one ribosome being tied to the synthesis of just one protein, it could travel along one message, synthesizing one protein, and then go onto a further messenger RNA, where it would synthesize a different protein. The PaJaMo results were easily explained by assuming that the messenger RNA was used only a few times before being destroyed. (We thought at first that it might be used only once but soon saw that this was unnecessarily restrictive.) This explained the linear increase of protein with time, since the messenger RNA for β-galactosidase soon reached an equilibrium concentration in which messenger synthesis was balanced by messenger degradation. This sounded wasteful but it allowed the cell to adjust rapidly to changes in its environment.

That evening I held a party at the Golden Helix. We often had parties (the molecular biologists' parties were considered to be the liveliest in Cambridge), but this one was different. Half the guests, such as the virologist Roy Markham, who had not been at the morning meeting, were just having a good time. The other half, in small groups, were earnestly discussing the new idea, seeing how it easily explained puzzling data and actively planning radically new key experiments to put the hypothesis to the test. Some of these were done later by Sydney, on a visit to Cal Tech, with François and Matt Meselson.

It is difficult to convey two things. One is the sudden flash of enlightenment when the idea was first glimpsed. It was so memorable that I can recall just where Sydney, François, and I were sitting in the room when it happened. The other is the way it cleared away so many of our difficulties. Just a single wrong assumption (that the ribosomal RNA was the messenger RNA) had completely messed up our thinking, so that it appeared as if we were wandering in a dense fog. I woke up that morning with only a set of confused ideas about the overall control of protein synthesis. When I went to bed all our difficulties had resolved and the shining answers stood clearly before us. Of course, it would take months and years of work to establish these new ideas, but we no longer felt lost in the jungle. We could survey the open plain and clearly see the mountains in the distance.

The new ideas opened the way for some of the key experiments used to crack the genetic code, since one could now conceive *adding* special messengers to ribosomes (either natural messengers or synthetic ones), an idea that before had no meaning.

Naturally you may be compelled to ask: Why had we not seen it before? In a certain sense we had, but because nothing seemed to support it we had not recognized it as important. It required us to accept, first, that the RNA we *did* see in the cytoplasm was *not* the messenger RNA and therefore had some other function. Exactly what that function is, is not clear even to this day, though we can make educated guesses. It also required us to hypothesize a key species of RNA *that had never been seen.* I wish I had been bold enough to take this step, but my natural caution must have prevented me. The irony was, of course, it had been seen in one particular case (the phage-infected cell) but we had not recognized it till that fateful Good Friday morning. Of course, messenger RNA was bound to be discovered eventually, but there is little doubt in my mind that this revelation speeded up the process considerably. After that, it might be said, the experiments wrote themselves. Nothing remained but the hard work: a happy state of affairs.

12

Triplets

ALTHOUGH SYDNEY AND I clearly realized that the genetic code was a biochemical problem, we still had hopes that genetic methods could contribute to the solution, especially as genetic methods, using the right material, can be very fast, whereas biochemical methods are often rather slower. Seymour Benzer had used genetic methods to demonstrate that the genetic material was almost certainly one-dimensional. The question had been inspired by the DNA double helix but the method used was entirely original.

In order to map a gene very finely, it is necessary to pick up rather rare individuals. The nearer two mutants are in a gene, the rarer will be the act of genetic recombination between them. Benzer had chosen a system with two advantages. The genes in question were in the bacteriophage T4, a virus that attacked and killed the cells of *E. coli.* The virus grows fast and recombines at a high rate. He had chosen the gene called r_{II}—actually a pair of genes next to each other—because it had a remarkable technical advantage. By using appropriate strains of the host cell, it was possible to pick up one virus having a wild-type gene even if mixed with millions of viruses with the mutant version. Thus very rare recombinant genes could be detected, so rare that Benzer calculated that even adjacent base pairs on the DNA could be separated. Unfortunately there was no corresponding method for picking up a mutant among a vast excess of wild type, but, on the appropriate host the plaque—the little colony formed by the growth of one bacterium in the lawn of *E. coli* in the petri dish—looked different and was easy to recognize. A single mutant plaque in a petri dish with several hundred wild-type plaques could be spotted fairly easily.

The conventional way to map would have been to pick a series of distinct mutants and then find the recombinant distance between any pair of them. More elaborate methods using three mutants were also possible, but all these involved the counting of hundreds and thousands of plaques, which was very laborious.

Benzer, always one to avoid unnecessary work, thought of a better method. As well as point mutations, he found that some of his mutants appeared to be deletions. They mapped as lines on his genetic map, since they appeared to overlap at least two of the point mutations. He was thus able to collect a whole series of deletions. If two deletions overlapped then genetic recombination could never give back the intact wild type, since the overlapping portion was in neither parent and thus could not be regained. On the other hand, if the two deletions did *not* overlap, then an appropriate recombination event could restore the wild type.

An analogy may help to make this clearer. Imagine two defective copies of a book, one with pages 100 to 120 missing and the other with pages 200 to 215 missing. Clearly from these two copies, each with a single, contiguous deletion, we could recover the text of the complete book. However, if the second book, instead of having pages 200 to 215 missing, lacked pages 110 to 125, there would then be no way of recovering pages 110 to 120, since they would be missing in both copies.

To make the analogy closer we have to extend it a little. Imagine that the book contained very detailed instructions for making a complicated instrument. Assume also that if any one page were missing, then either no instrument would be made or, if it were, it would be a dud one. Finally, assume that we had millions of copies of each of the defective books. The rule then was: Select one copy of each type of book. Take the first *n* pages from one book and the remaining pages from the other. See if this new hybrid book will produce an instrument that works. Do this a million times, selecting the cross-over page (page *n*) at random each time. If occasionally a good instrument was produced, the two deletions did *not* overlap. If a good instrument was *never* produced, then the deletions probably overlapped.

This may seem an elaborate way of doing it, but since we could not look inside the phage this was the only method we had. All Benzer had to do, then, was to mate the two viruses together, by simultaneously infecting a culture of *E. coli* with them. After they had grown and recombined inside the bacteria, the viruses could be plated on a petri dish having the special host strain. If the deletions did not overlap, then there would be some recombinant plaques on the plate. If they *did* overlap, there would be none. There was no need for laborious counting. All that was needed was a simple yes or no answer.

Benzer argued that if the gene was two-dimensional then eventually he should find a special pattern with four deletions. Deletion *A* would overlap *B* and *C*, as would *D*; but deletions *B* and *C* would not overlap nor would *A* and *D*. (See figure 12.1.) It is easy to see that this cannot happen if the gene is one-dimensional. Benzer picked up hundreds of deletions and crossed them all against one another in pairs. The situation shown in the figure never arose. Therefore, he concluded, the gene is probably one-dimensional. His results also allowed him to put all his deletions in order, so he would see roughly where each one was in the map of the gene.

For a variety of reasons, our group had chosen the same system

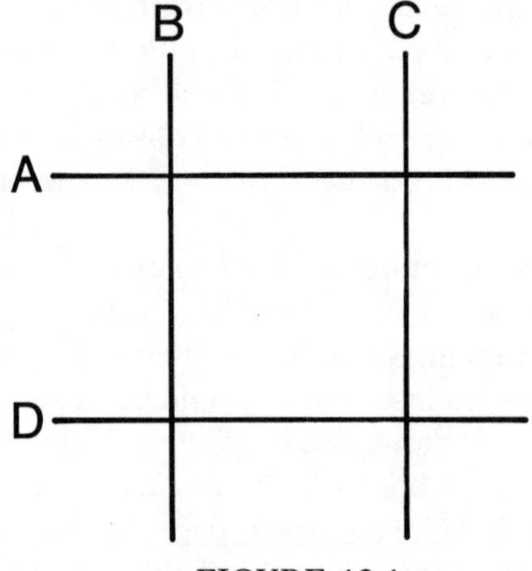

FIGURE 12.1

To show that in *two* dimensions we can have A overlapping both B and C, and D also overlapping B and C, without A overlapping D or B overlapping C. This is impossible if A, B, C, and D are segments of a (one-dimensional) straight line. Since Benzer never found this pattern of overlapping among his many deletions, he concluded, correctly, that the gene he was studying was one-dimensional. This was compatible with it being made of DNA.

to work on. Our main interest was in the different type of mutants produced by different chemicals and also in the reverse mutations these chemicals produced. Mutants appeared to fall broadly into two classes. Most chemicals produced mutants of the first class. However, the mutants produced by chemicals of the acridine type fell into the other. Each class was most easily reverted by the type of mutagen that produced it. Ernst Freese had suggested that one class corresponded to transitions (purine to purine, or pyrimidine to pyrimidine—see appendix A) while the other corresponded to transversions, as they were called (purine to pyrimidine, or vice versa). We had come up with another idea. Some mutants were leaky—that is, they showed the gene was active to some extent, though of course not fully active—whereas others were nonleaky—that is, had essentially no activity. We noticed that mutants produced by proflavin (a typical acridine) were almost always nonleaky. This led us to suggest that the proflavin mutants were tiny deletions or additions to the base sequence, whereas all the other class of mutants were base substitutions of one sort or another. However, we lacked further evidence to confirm this idea.

Meanwhile I had come up with a quite different idea. Pondering over how an RNA molecule could act as a message, I wondered if it could fold back on itself, thus forming a loose double-helical structure. The idea was that some bases could pair whereas others, which did not match according to the pairing rules, would loop out. The "code" would then depend either on the looped-out bases or the paired ones, or some more elaborate combination of these two possibilities. The idea was really rather vague, but it made one important prediction. A mutant at one end of the message might, in theory, be capable of being compensated in its effect by another, toward the other end, that paired with it. Thus some mutants should have distant "suppressors," as they are called, within the same gene.

I rather liked this idea, but nobody else thought very much of it. Up to that time I had not done any phage genetics myself, being content to look at the experimental results of my colleagues. Since nobody seemed keen to test my idea, I decided to test it myself. It was not difficult to learn how to do phage genetics, especially with expert help at hand. In spite of this I made some elementary mistakes, fortunately soon corrected. The experiments also taught me how superficial my knowledge was, even though I had taken part in numerous discussions about this very system. There is nothing like actively doing experiments to make one realize all the ins and outs of a technique. It also helps to fix the details in one's head, especially as reading the "experimental methods" section of most scientific papers is more boring than almost anything I know.

Naturally I chose the r_{II} genes for the experiment, concentrating on the second one, the so-called B cistron. (Cistron was Benzer's fancy name for a gene, as defined by the so-called cis-trans test.) I selected a mutant from our stock, tried to find a revertant—one more like the wild type—and then looked to see if this reversion was due to a second mutation somewhere else in the same gene. If I couldn't find it I went on and tried again with another mutant.

At first I couldn't find any suppressors. Presumably the change that reverted the mutant to wild type was at, or very close, to the original change, too close for me to pick it up. Leslie Orgel came over to coffee one day. As he looked over my shoulder I told him what I was doing and that so far I had no results. He went off to join the others while I quickly scored the remaining plates. To my delight I found I had a candidate suppressor.

Before long I had three mutants with suppressors, fortunately spaced out along the map. I isolated the suppressors and proceeded to map them. My theory was immediately refuted. Instead of each suppressor mapping at the predicted place, some distance away on the map, each had its suppressor quite near it. The suppressor effect must be due to some other reason.

Unbeknown to me, other people had also noted that a mutant in r_{II} could have a suppressor in the same gene. Perhaps the most striking example occurred at Cal Tech. Dick Feynman, the theoretical physicist, had become sufficiently interested in these genetic problems that he had decided to do some experiments himself. He stumbled upon an example of an internal suppressor. Not knowing what it might imply, he asked his mentor, Max Delbrück. Max suggested that the original mutant had produced a changed amino acid and that the second mutant had changed another amino acid, elsewhere in the protein, which somehow compensated for the first change. It was easy to see that this might happen, but one would not expect it to be too common.

I was certainly aware of this possibility but I was not happy with it, partly because I had a very detailed knowledge of what little was then known about protein structure. I decided to try to see how many different suppressors a particular mutant might have. I had to select one of my three to study further and, sensibly, I chose the one whose suppressor was the least close to the parent mutant, hoping this would give me more elbow room. I also noted that two of my three mutants had been produced by proflavin. Even though this was hardly statistically significant, by any test, it seemed suggestive to me.

By now I was a little more experienced, so the experiments went fairly quickly. Phage genetics has the advantage that experiments are rather fast, once everything is set up. It does not take long to carry out a hundred crosses, since the manipulations are easy and an actual cross takes only about twenty minutes, this being the time for the phage to infect the bacterium, to multiply inside it (exchanging genetic material in the process), and to burst open, thus killing the cell. The results of the cross must then be plated out on petri dishes, to which a thin film of bacteria has been added. Then the dishes have to be incubated, to produce a lawn of bacteria. Where a single phage has landed and infected a cell, a colony of phage will grow, killing the local bacteria as it does so, forming a clear little hole (called a plaque) in the lawn of growing bacteria on the surface of the plate. This process takes a few hours, so one has a brief respite while it is going on. Then the petri dishes have to be taken from the 37 ° C incubator and examined to see whether they have plaques or not and, if so, of what type. Interesting plaques are then "picked"—that is, a few phage are picked up with a little piece of paper or a toothpick; grown further; and the process repeated a second time to make sure the phage stock is a pure one. If one works reasonably hard it is possible to complete one extra set of crosses in one day, and prepare for a new set the next day.

As the experiments got more interesting I found that, with careful planning, I could get through *two* successive sets of crosses in one day. This involved starting promptly in the morning, going home for lunch, more experiments in the afternoon, home for dinner, and a final set after dinner. Fortunately Odile and I then lived within a few minutes of the laboratory, an easy walk through the center of historic Cambridge, so I did not find the work unpleasant. In fact, Odile has told me she had never seen me so cheerful as during the period when I did experiments all the time, but this may have been partly because, for weeks on end, all the experiments seemed to work perfectly.

I soon found that my initial mutant had not one, but several, distinct suppressors, all of which mapped fairly close to the original mutant. I decided that I would have to call them all by a distinctive name. I often worked through the weekend, taking Monday off so that our laboratory kitchen (which did all the washing up and also prepared petri dishes for our use) could catch up. It happened that it was a weekend when I needed a new name, and nobody else was around. Mutants were usually called by a letter, followed by a number. Thus P31 meant the thirty-first mutant in the P series, probably produced by proflavin. Unfortunately I could not remember for certain which letters had already been used, so I decided to rename my mutant FC0, since I was quite sure that no one had used my initials to name mutants. The new suppressives were then named FC1, FC2, and so on. This use of my own initials suggested to some people that I must be conceited, but the real explanation was that I have a rather fallible memory.

The new suppressors all seemed like good, nonleaky mutants. So why not, I argued, see if they too had suppressors? And indeed they had. I even went a step further and found suppressors of suppressors of suppressors.

So what was going on? Fortunately we had the right ideas already at hand. Assume that the genetic message was read (to produce a protein) in steps of three bases at a time, starting from one particular point in the message. To make it clearer, let us take an extremely simple message that merely repeats the triplet TAG over and over again:

. . . TAG, TAG, TAG, TAG, TAG, TAG, . . .

the dots indicating that there is message both before and after such a sequence. Commas have been added to show in which "phase" the sequence has to be read. I assumed that this phase was determined by a special "start" signal, somewhere to the left of the stretch shown.

Assume that our original mutant (now called FC0) had added a base to the base sequence. Then, from that point on, the reading would be out of step (out of phase) and thus would produce a nonsense protein, a protein whose amino acid sequence, following the mutant, was completely incorrect, so that the gene product could not function.

Our simple sequence might have become

$$\ldots \text{TAG, T}\overset{+}{\text{C}}\text{A, GTA, GTA, GTA, GTA,} \ldots$$
$$\text{correct}\rightarrow \longleftarrow\text{————— incorrect —————}\longrightarrow$$

(The added base has been shown, for clarity, as a C, but it could have been any of the four bases.)

Then, on this interpretation, a suppressor, such as FC1, was the *deletion* of one base at a point nearby. In between FC0 and FC1 the message would still be incorrect, being read in the wrong phase, but elsewhere the reading would be normal.

Our example might thus become:

$$\ldots \text{TAG, T}\overset{+}{\text{C}}\text{A, GTA, GT}\overset{-}{\not{\text{A}}}\text{G, TAG, TAG,} \ldots$$

that is, $\ldots \text{TAG, T}\overset{+}{\text{C}}\text{A, GTA, G}\overset{-}{\text{T}}\text{G, TAG, TAG,} \ldots$

$$\text{correct}\rightarrow \leftarrow \text{incorrect} \longrightarrow \leftarrow \text{correct} \longrightarrow$$

If the altered bit of the amino acid sequence was not crucial (and in this case there was other evidence to suggest this), then the protein would still function fairly well and the double mutant (FC0 + FC1) would behave more like a wild type than like a nonleaky mutant.

I therefore labeled all the first set of suppressors −. The next set, the suppressors of the first set of suppressors, we labeled +, and *their* suppressors we labeled −.

I had started these experiments early in May and by now summer was advancing. I had previously arranged to take my family on a summer holiday, almost the first proper holiday we had ever had, since by now my financial position was a little easier. We had rented, for a very small sum, a large villa on the old mountain in Tangier, a town in North Africa, just opposite Gibraltar. Here we lived in splendor, with one Arab servant living in and another coming each day. Odile and our German au pair girl, Eleanora, learned how to shop for food in the Arab market, bargaining, walking away, and so on. Our two daughters improved their swimming on the beach while I usually spent the day on the terrace, in the dappled shade of the palm trees.

On the way to Tangier I attended a scientific meeting. Even in those days scientists were reluctant to go to a meeting unless it was in some interesting place. This meeting was at Col de Voz halfway up Mont Blanc. I reported my preliminary results, which eventually were published as a very brief communication related to the meeting.

After a month in Tangier I went off to the 1961 Biochemical Congress at Moscow, leaving my family to stay at the villa for another week or so. Moscow then was very different from my first visit in 1945, during the war. Now it was summer, rather than the depth of winter, and everything was brighter and more prosperous than in the drab days of wartime. I stayed in a student's room in the university, where the meeting was held, and got to know some of our Russian hosts. A dominant figure was Igor Tamm, the Russian physicist. The influence of Lysenko, the man who had, for a period, killed genetics in the USSR, was very much on the wane. I sensed that his eclipse was largely the work of physicists like Tamm who had considerable political influence and who could recognize scientific nonsense when they saw it. A number of us were invited to give talks to the biological section of the Russian Atomic Energy Research establishment, something that could not have happened a few years before. We gave our talks in English, but they were brilliantly translated (in chunks, as we went along) by Bressler, a Russian scientist we had already met when he had visited Cambridge. Bressler not only understood what we were saying but in some cases, as I could tell by listening to him, filled out the "references" the speakers were giving, a truly remarkable performance.

The Moscow meeting was made especially interesting because of the results reported by Marshall Nirenberg, then almost unknown. I had heard rumors of these experiments but no details. Matt Meselson, whom I ran into in a corridor, alerted me to Marshall's talk in a remote seminar room. I was so impressed that I asked Marshall to take part in a much larger meeting, of which I was the chairman. What he had discovered was that he could add an artificial message to a test-tube system that synthesized proteins and get it to direct some synthesis. In detail, he had added poly U—the RNA message consisting entirely of a sequence of uracils—to the system and it had synthesized polyphenylalanine. This suggested that UUU (assuming a triplet code) was a codon for phenylalanine (one of the "magic twenty" amino acids), as indeed it is. I later claimed that the audience was "startled" (I think I originally wrote "electrified") to receive this news. Seymour Benzer countered this with a photograph showing everyone looking extremely bored! Nevertheless it was an epoch-making discovery, after which there was no looking back.

There was also a measure of social life during the week in Moscow. I enjoyed visiting an old-style apartment, with heavy furniture and a bed behind a large bookcase. Also a more modern one, with a much lighter tone. The owner collected modern Russian art. I was amused to notice Alex Rich demonstrating a strange new American dance to our host, a dance I later recognized as the twist. As Alex's waist is not very pronounced, the twist, as demonstrated by him, was somewhat less than free-flowing.

I returned to Cambridge. The next step was to do further experiments to validate the ideas that there was some sense in labeling each of our new r_{II} mutants as either + or −. The theory predicted that any combination of the type (+ +) or (− −) would be a mutant. My colleagues and I constructed quite a number of such pairs and they were all nonleaky mutants, as predicted. The simple theory also predicted that *any* combination of the type (+ −) would be wild type, or approximately so. Of course we knew this to be true in some cases, since that was how we had picked up the suppressor in the first place, but many other combinations (of a + with −) had never been tested. These we called "Uncles and Aunts," since creating them often involved putting together a mutant of one generation with a mutant from a previous generation, but one other than the one it was descended from. I had asked Sydney to see that some of these were tried while I was away but he had other ideas, so I had to do it myself when I returned.

At this point a small complication arose. Some of the (+ −) combinations, predicted to be wild type, turned out to be mutant. We explained these away by assuming that in some cases the small local phase shift between the + and the − produced a "nonsense" mutant. We know now that these nonsense sites were due to a triplet that terminated the polypeptide chain, thus producing a nonfunctional protein fragment. I also realized that this depended on the precise phase of the reading. For a nonoverlapping triplet code there is one correct phase but *two* incorrect ones, so that a combination (+ −), that is, + followed by −, will be locally different from a (− +) combination.

To return to our simple example, a (+ −) combination might be:

. . . TAG, T$\overset{+}{\text{C}}$A, GTA, GTA, G$\overset{-}{\text{T}}$G, TAG, . . .
correct→ ←—— incorrect ——→ ←correct

and a (− +) combination

. . . TAG, T$\overset{-}{\text{G}}$T, AGT, AGT, AG$\overset{+}{\text{C}}$, TAG, . . .
correct→ ←—— incorrect ——→ ←correct

The first has GTA between the two alterations; the second has AGT. We showed that our (+ −) or (− +) failures obeyed this rule, which made us fairly confident our ideas were along the right lines.

Previous to this Sydney had an idea. He reasoned that a (+ +) mutant might backmutate to a wild type. He tried one, but the back mutation must have been too close to an existing one, since he could not separate it. Another, slightly more laborious approach was to construct a triple mutant, of the form (+ + +) or (− − −). According to our ideas, these should be wild type, since the three successive changes in phase should have restored the correct phase, always assuming, of course, that it was a triplet code.

For our simple sequence, an example might be

. . . TAG, TA$\overset{+}{\text{C}}$, GTA, G$\overset{+}{\text{C}}$T, AGT, AG$\overset{+}{\text{C}}$, TAG, . . .
correct→ ←——— incorrect ———→ ←correct

A direct but laborious way of constructing such a triple mutant is to choose three mutants, not too far apart and all +, then to construct two pairs, each of which has the middle mutant in common. (See figure 12.2.) This is the laborious part, since there is no way to select for such a combination of mutants. One has to do the cross and laboriously test the offspring having a mutant phenotype, by taking each one apart, till one finds one which is indeed the

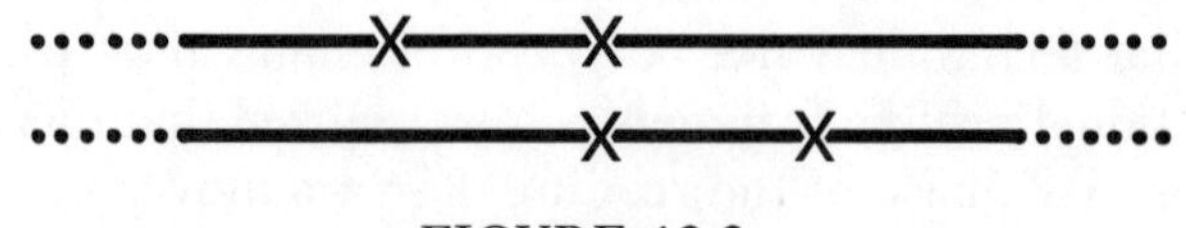

FIGURE 12.2

Each line represents one of the two parental strands. Each X represents a mutation. It is impossible to recombine the two parental strands to give a strand having no mutations at all. The middle mutation will always be there. Moreover, some of the progeny may have all three deletions on the same strand.

(+ +) one is looking for. The final step is easy. One simply crosses two doubles together. Since each contains the middle mutant of the three, there is no way that the cross can produce true wild type. If apparently wild-type plaques do arise from the cross, they are highly likely to be the sought-after (+ + +). In any case it is then very easy to check that this is so by taking the presumed triplet apart.

Of course, the triplet would look like a wild type only if the code was a triplet code. If the bases were read four or five at a time, which for all we knew was not impossible, the (+ + +) would be a mutant, and we would have to construct a (+ + + +) or even a (+++++). Not everybody in the lab believed the experiment would work. I was almost certain it would. So was Sydney, who was away at the time in Paris. He had listed three possible (+++) combinations to try, but after he had left I fortunately realized that two of them would probably not work because they would produce a chain terminator, so we constructed the third one that was likely to be free from this complication.

By this time I had co-opted Leslie Barnett to help me. The final crosses were duly carried out and the pile of petri dishes put in the incubator. We came back after dinner to inspect them. One glance at the crucial plate was sufficient. There were plaques in it! The triple mutant was showing the wild-type behavior (phenotype). Carefully we double-checked the numbers on the petri dishes to make sure we had looked at the correct plate. Everything was in order. I looked across at Leslie. "Do you realize," I said, "that you and I are the only people in the world who *know* it's a triplet code?"

The result, after all, was remarkable. Here we had three distinct mutants, any one of which knocked out the function of the gene. From them we could construct the three possible double mutants. Each one of these also made the gene nonfunctional. Yet if we put all three together in the same gene (and we did separate experiments to show that they had to be in the same virus, not some in one and the rest in another separate virus), then the gene started to function again. This was easy to understand if the mutants were indeed additions or deletions and if the code was indeed a triplet one. In short, we had provided the first convincing evidence that the code was a triplet code.

I exaggerate slightly. The evidence would also fit a code with *six* bases in each codon, but this possibility, as subsidiary experiments showed, was very unlikely and hardly to be taken seriously.

There still remained a lot of work to fill out our results. We constructed not one but six distinct triples—five of the (+ + +) type and a single (− − −) one—and showed they all behaved like the wild type. I was even busier than before, though by now Leslie was giving me a lot of help. Not that there were not distractions. One evening, after dinner, I was working away in the lab when a glamorous friend of mine turned up and stood behind me while I continued to manipulate the tubes and plates. "Come to a party," she said, running her fingers through my hair. "I'm far too busy," I said, "but where is it?" "Well," she said, "we thought we'd hold it in *your* house." Eventually a compromise was reached. She and Odile would organize a small party and I would join them when I'd finished.

Looking back, it seems remarkable how little we worked—I was away for about six weeks in the summer, on my trips to Mont Blanc, Tangier, and Moscow—and yet how hard we worked and how fast. I had started the key experiment early in May. Yet the paper was published in *Nature* in the last issue of the year.

We didn't stop there. Sydney in particular did many further ingenious experiments with the system. Eventually we decided we had better publish a really full account of it, so Leslie Barnett and I worked hard at tidying up all the loose ends. This had one remarkable result. It was known by then that the two triplets UAA and UAG were chain terminators. I was convinced that UGA was a third one. Sydney had devised a complicated way of testing this genetically, but the experiments always told us that it wasn't. When we came to write our results up, we noticed that not all the possible experiments of this type had been done. Rather than have a gap in one of our tables, we asked Leslie, as a matter of routine, to do the ones that had been overlooked. To our surprise, the experiment now worked! We then repeated all the earlier ones, and this time they worked too! It transpired that when they were first done, we had included a set of controls to make sure everything was as it should be. Unfortunately, in each experiment one control or another had been skipped. When all the controls worked correctly, the experiment suggested strongly that UGA was a chain terminator.

We had planned to give our results a decent burial in the august pages of the *Philosophical Transactions of the Royal Society.* As we now had a result of some interest, we took the experiments out of the proposed *Philosophical Transactions* paper and made them into a separate paper that appeared shortly afterward in *Nature.* I was somewhat surprised to find my name on the draft paper, since the convention in our lab was that one did not put one's name on a paper unless one had made a significant contribution to it. Mere friendly advice was not enough. "Why," I asked Sydney, "have you added my name?" He grinned at me. "For persistent nagging," he said, so I let it stand.

One of the more laborious experiments that Leslie did was to put *six* +'s together in one gene and show that the result was like wild type. It is difficult to convey just how tedious and complicated such an experiment is. The required (+ + + + + +) must be put together in stages, testing at each stage to see that the gene does indeed have the structure it is supposed to. When the final combination has been produced and tested, it must still be taken apart, step by step, to make sure that it was what we thought it was. Even an outline description of all Leslie did took up several of the large pages of *Philosophical Transactions.*

When we were going through the final manuscript, I told Sydney that I supposed he and I would be the only people in the world who would ever read through it carefully. For fun we decided to add a fake reference, so at one point we put "Leonardo da Vinci (personal communication)" and submitted it to the Royal Society. One (unknown) referee passed it without comment, but we had a phone call from Bill Hayes, the other referee, who said, "Who's this young Italian working in your lab?" so reluctantly we had to take it out.

The demonstration by genetic methods that the code was a triplet code was a tour de force, but in only a short time it was established by direct biochemical methods. Of more importance in the long run was the demonstration that acridine mutants caused small deletions and omissions. Even this was not unsuspected, since Leonard Lerman had produced very suggestive physical chemical evidence that acridines slipped *in between* the bases of DNA, and this could easily lead to additions or deletions of DNA when it was copied. Moreover, the theory had to be firmly established by direct biochemical methods. Both the biochemists Bill Dreyer and George Streisinger planned to do this though they were somewhat slow in getting the answer—at that time it was technically difficult to do the biochemistry. Each month or so Sydney and I would debate whether we should tackle it ourselves but we were reluctant to do this, especially as George was an "old boy"—meaning he had spent some time in our lab. Eventually George got the answer, working not on the unknown products of the two genes but on phage lysozyme. It came out exactly as we expected. In between the mutants a string of amino acids was indeed altered, and moreover, they fit well with what was known of the genetic code, which was just coming out.

A little later I was at a meeting at the Villa Serbelloni on Lake Como, organized by the biologist Conrad Waddington (always called Wad by his friends). There for the first time I met the mathematician René Thom. Almost the first thing he told me was that our work on the acridine mutants must be wrong. As I had just heard that our ideas had been confirmed biochemically, I was somewhat surprised and asked him why he thought so. He explained that if one made, say, a triple mutant, one necessarily got a Poisson distribution of single, double, quadruple, and so on, and so our arguments were not sound. Since we had laboriously put together our multiple mutants (and tested each carefully), I saw immediately that his objection had no force, being based on a misunderstanding. Either he had not read our paper carefully enough or, if he had read it, he had not understood it. But then in my experience most mathematicians are intellectually lazy and especially dislike reading experimental papers.

My impression of René Thom was of a good mathematician but a somewhat arrogant one, who disliked having to explain his ideas in terms nonmathematicians could understand. Fortunately another topologist, Christopher Zeeman, also at the meeting, was exceptionally good at putting over Thom's ideas.

My other impression was that Thom really understood very little about how science was done. What he did understand he didn't like, and referred to it disparagingly as "Anglo-Saxon." He seemed to me to have very strong biological intuitions but unfortunately of negative sign. I suspected that any biological idea he might have would probably be wrong.

13

Conclusions

THE TIME HAS COME to try to pull all the threads together. In the episodes sketched earlier I have tried to suggest some aspects of biological research, both to illustrate its special character and also, by the way, to paint in a few glimpses of research as a human activity.

What gives biological research its special flavor is the long-continued operation of natural selection. Every organism, every cell, and all the larger biochemical molecules are the end result of a long intricate process, often stretching back several billion years. This makes biology a very different kind of subject from physics. Physics, either in its more basic forms, such as the study of the fundamental particles and their interactions, or in its more applied branches, such as geophysics or astronomy, is very different from biology. It is true that in the latter two branches we have to deal with changes over comparable periods of time and what we see may be the end result of a long historical process. The layers upon layers of rock exposed in the Grand Canyon would be an example. However, while stars may "evolve," they do not evolve by natural selection. Outside biology, we do not see the process of exact geometrical replication, which, together with the replication of mutants, leads to rare events becoming common. Even if we may occasionally glimpse an approximation to such a process, it certainly does not happen over and over again, till complexity is added to complexity.

Another key feature of biology is the existence of many identical examples of complex structures. Of course, many stars must be broadly similar to each other. Many crystals in geological rocks must have a basically similar structure. But in neither case do we find masses of stars or crystals that are identical in many small details. One type of protein molecule, on the other hand, usually exists in many absolutely identical copies. If this were produced by chance alone, without the aid of natural selection, it would be regarded as almost infinitely improbable.

Physics is also different because its results can be expressed in powerful, deep, and often counterintuitive general laws. There is really nothing in biology that corresponds to special and general relativity, or quantum electrodynamics, or even such simple conservation laws as those of Newtonian mechanics: the conservation of energy, of momentum, and of angular momentum. Biology has its "laws," such as those of Mendelian genetics, but they are often only rather broad generalizations, with significant exceptions to them. The laws of physics, it is believed, are the same everywhere in the universe. This is unlikely to be true of biology. We have no idea how similar extraterrestrial biology (if it exists) is to our own. We may certainly consider it likely that it too will be governed by natural selection, or something rather like it, but even this is only a plausible guess.

What is found in biology is *mechanisms,* mechanisms built with chemical components and that are often modified by other, later, mechanisms added to the earlier ones. While Occam's razor is a useful tool in the physical sciences, it can be a very dangerous implement in biology. It is thus very rash to use simplicity and elegance as a guide in biological research. While DNA could be claimed to be both simple and elegant, it must be remembered that DNA almost certainly originated fairly close to the origin of life when things were necessarily simple or they could not have got going.

Biologists must constantly keep in mind that what they see was not designed, but rather evolved. It might be thought, therefore, that evolutionary arguments would play a large part in guiding biological research, but this is far from the case. It is difficult enough to study what is happening now. To try to figure out exactly what happened in evolution is even more difficult. Thus evolutionary arguments can usefully be used as *hints* to suggest possible lines of research, but it is highly dangerous to trust them too much. It is all too easy to make mistaken inferences unless the process involved is already very well understood.

All this may make it very difficult for physicists to adapt to most biological research. Physicists are all too apt to look for the wrong sorts of generalizations, to concoct theoretical models that are too neat, too powerful, and too clean. Not surprisingly, these seldom fit well with the data. To produce a really good biological theory one must try to see through the clutter produced by evolution to the basic mechanisms lying beneath them, realizing that they are likely to be overlaid by other, secondary mechanisms. What seems to physicists to be a hopelessly complicated process may have been what nature found simplest, because nature could only build on what was already there.

The genetic code is a very good example of what I mean. Who could possibly invent such a complex allocation of the sixty-four triplets (see appendix B)? Surely the comma-free code (page 99) was all that a theory should be. An elegant solution based on very simple assumptions—yet completely wrong. Even so, there is a simplicity of a sort in the genetic code. The codons all have just three bases. The Morse code, by contrast, has symbols of different lengths, the shorter ones coding the more frequent letters. This allows the code to be more efficient, but such a property may have been too difficult for nature to evolve at that early time. Arguments about "efficiency" are thus almost always to be mistrusted in biology since we don't know the exact problems faced by myriads of organisms in evolution. And without knowing that, how can we decide what form of efficiency paid off?

There is a more general lesson to be drawn from the example of the genetic code. This is that, in biology, some problems are not suitable or not ripe for a theoretical attack for two broad reasons. The first I have already sketched—the current mechanisms may be partly the result of historical accident. The other is that the "computations" involved may be exceedingly complicated. This appears to be true of the protein-folding problem.

Nature performs these folding "calculations" effortlessly, accurately, and in parallel, a combination we cannot hope to imitate exactly. Moreover, evolution will have found good strategies for exploring many of the possible structures in such a way that short-cuts can be taken on the paths to the correct fold. The final structure is a delicate balance between two large numbers, the energy of attraction between the atoms, and the energy of repulsion. Each of these is very difficult to calculate accurately, yet to estimate the free energy of any possible structure we have to estimate their difference. The fact that it usually happens in aqueous solution, so that we have to allow for the many water molecules bordering the protein, makes the problem even more difficult.

These difficulties do not mean we should not look for the broad principles involved (for example, a protein that exists in aqueous solution folds to keep many of its water-hating side groups out of contact with the water), but it does mean that it may be better to try to go around such problems and not try to tackle them head on at too early a stage.

A number of other lessons can be drawn from the history of molecular biology, though it would be easy to find examples in other branches of science as well. It is astonishing how one simple incorrect idea can envelop the subject in a dense fog. My mistake in thinking that each of the bases of DNA existed in at least two different forms is one such case. Another, more dramatic in some ways, was the assumption that the ribosomal RNA was the messenger RNA. And yet see how plausible this mistaken idea was. Jean Brachet, the embryologist, had shown that cells with a high rate of protein synthesis had large amounts of RNA in their cytoplasm. Sydney and I knew there had to be a messenger to convey the genetic message of each gene from the DNA in the nucleus to the ribosomes in the cytoplasm, and we assumed that this had to be RNA. In this we were right. Who would have been so bold as to say that the RNA we saw there was *not* the messenger but that the messenger was another kind of RNA, as yet undetected, turning over rapidly and thus probably there in small amounts? Only the gradual accumulation of experimental facts that appeared to contradict our base idea could jolt us out of our preconception. Yet we were acutely aware that something was wrong and were continually trying to find out what it was. It was this dissatisfaction with our ideas that made it possible for us to spot where the mistake was. If we had not been so conscientious in dwelling on these contradictions we should never have seen the answer. Eventually, of course, someone else would have spotted it, but the subject would have advanced less rapidly—and we would have looked very silly.

It is not easy to convey, unless one has experienced it, the dramatic feeling of sudden enlightenment that floods the mind when the right idea finally clicks into place. One immediately sees how many previously puzzling facts are neatly explained by the new hypothesis. One could kick oneself for not having the idea earlier, it now seems so obvious. Yet before, everything was in a fog. Often it becomes clear that to prove the new idea a different sort of experiment is needed. Sometimes these experiments can be carried out in a remarkably short time and, if successful, serve to put the hypothesis beyond reasonable doubt. On such occasions one can go from muddled puzzlement to virtual certainty in the space of a year or even less.

I have discussed earlier (in chapter 10) the importance of general, negative hypotheses (if one can find good ones), the mistake of mixing up a process with the rather different mechanisms that control it, and especially the importance of not mistaking a minor, subsidiary process for the main mechanism one is interested in. However, the principal error I see in most current theoretical work is that of imagining that a theory is really a good model for a particular natural mechanism rather than being merely a demonstration—a "don't worry" theory. Theorists almost always become too fond of their own ideas, often simply by living with them for so long. It is difficult to believe that one's cherished theory, which really works rather nicely in some respects, may be completely false.

The basic trouble is that nature is so complex that many quite different theories can go some way to explaining the results. If elegance and simplicity are, in biology, dangerous guides to the correct answer, what constraints can be used as a guide through the jungle of possible theories? It seems to me that the only really useful constraints are contained in the experimental evidence. Even this information is not without its hazards since, as we have seen, experimental facts are often misleading or even plain wrong. It is thus not sufficient to have a rough acquaintance with the experimental evidence, but rather a deep and critical knowledge of many different types of evidence is required, since one never knows what type of fact is likely to give the game away.

It seems to me that very few theoretical biologists adopt this approach. When confronted with what appears to be a difficulty, they usually prefer to tinker with their theory rather than seeking for some crucial test. One should ask: What is the essence of the type of theory I have constructed, and how can that be tested? even if it requires some new experimental method to do so.

Theorists in biology should realize that it is extremely unlikely that they will produce a useful theory (as opposed to a mere demonstration) just by having a bright idea distantly related to what they imagine to be the facts. Even more unlikely is that they will produce a good theory at their first attempt. It is amateurs who have one big bright beautiful idea that they can never abandon. Professionals know that they have to produce theory after theory before they are likely to hit the jackpot. The very process of abandoning one theory for another gives them a degree of critical detachment that is almost essential if they are to succeed.

The job of theorists, especially in biology, is to suggest new experiments. A good theory makes not only predictions, but surprising predictions that then turn out to be true. (If its predictions appear obvious to experimentalists, why would they need a theory?) Theorists will often complain that experimentalists ignore their work. Let a theorist produce just one theory of the type sketched above and the world will jump to the conclusion (not always true) that he has special insight into difficult problems. He may then be embarrassed by the flood of problems he is asked to tackle by those very experimentalists who previously ignored him. If this book helps anyone to produce good biological theories, it will have performed one of its main functions.

14

Epilogue: My Later Years

IN JUNE 1966 the annual meeting at Cold Spring Harbor Laboratory was on the genetic code. It marked the end of classical molecular biology, because the detailed delineation of the genetic code—the little dictionary—showed that, in outline, the basic ideas of molecular biology were largely correct. It was remarkable to me and to most other people, both in the field and outside it, that we should have reached this point so quickly. When I started biological research in 1947 I had no suspicion that all the major questions that interested me—What is a gene made of? How is it replicated? How is it turned on and off? What does it do?—would be answered within my own scientific lifetime. I had selected a topic, or series of topics, that I had assumed would last out my active scientific career, and now I found myself with most of my ambitions satisfied.

Of course, not all these questions had received detailed answers. We still did not know the base sequence of any gene. Our ideas about the biochemistry of gene replication were oversimplified. Only in bacteria did we know how a gene was controlled, and even in this case the molecular details were lacking. About the control of genes in higher organisms we knew hardly anything. And although we knew that messenger RNA directed protein synthesis, the site of protein synthesis—the ribosome—was little more than a black box to us. Nevertheless by 1966 we realized that the *foundations* of molecular biology were now sufficiently firmly outlined that they could be used as a fairly secure basis for the prolonged task of filling in the many details.

Sydney Brenner and I thought that it was time to move on to new fields. We selected embryology—now often called by the more general term developmental biology. Sydney, after much reading and thinking, chose the little nematode worm *Caenorhabditus elegans* as a suitable organism to study, since it bred fast, was easy to grow in the lab, and had unusual but attractive genetics. (It is a self-fertilizing hermaphrodite.) Almost all the work now being done on this little animal—it is even used for studies on aging—springs from these pioneer studies of Sydney.

I decided that a key feature of development was gradients, whatever they were. In some way a cell in an epithelium (a sheet of cells) seemed to know where it was in the sheet. This was ascribed to the existence of "gradients" of some form or another—possibly the regular change in concentration of a chemical from one part of the sheet to another. The nature of these postulated gradients was then quite unknown. At about this stage Peter Lawrence joined us, and I followed closely his work on gradients in the cuticle of insects, which had been pioneered by Michael Locke. My colleagues Michael Wilcox and Graeme Mitchison studied an even simpler system, the pattern of cells in the long chains of cells formed by one of the blue-green algae (now called a bacterium). In spite of all their efforts, it proved impossible to get a foothold in the biochemical basis of the problem—what molecules were used to form this gradient or that?—and eventually I moved on to other aspects of the subject. I became interested in the histones, the small proteins found associated with DNA in the chromosomes of higher organisms, and attended closely to the work of my colleagues Roger Kornberg, Aaron Klug, and others, which led to the structure of nucleosomes, the small particles on which chromosomal DNA is wound.

In 1976 I decided to go to The Salk Institute on a sabbatical. The Salk (the full title is The Salk Institute for Biological Studies) is situated a little back from the cliffs overlooking the Pacific Ocean in La Jolla, a suburb of San Diego in lower Southern California. For twelve years, from shortly after the official start of the institute on

December 1, 1960, I had been a nonresident Fellow (effectively a member of a visiting committee), and indeed I had been involved with it even before it started. In the very early days "Bruno" Bronowski and I would fly from London to Paris to consult with Jonas Salk, Jacques Monod, Mel Cohn, and Ed Lennox on such fascinating topics as the by-laws for the proposed institute.

The president of the Salk Institute, Dr. Frederic de Hoffmann, went to great efforts to tempt me to stay on there. Eventually he persuaded the Kieckhefer Foundation to endow a chair for me. I resigned from the Medical Research Council. Odile and I took up residence in Southern California, where we have been ever since.

California is effectively bounded on the east by the desert, on the west by the Pacific, on the south by Mexico, and on the north by the state of Oregon, where it appears to rain much of the time. California is almost twice the size of Britain, has a little less than half the population of the U.K., and is appreciably more affluent. It has a large and impressive system of universities. Odile and I are resident aliens—immigrants, that is—though we remain British citizens. An immigrant doesn't have a vote but otherwise has all the privileges and duties of a U.S. citizen, including paying taxes.

Personally I feel at home in Southern California. I like the prosperity and the relaxed way of life. The easy access to the ocean, the mountains, and the desert is also an attraction. There are miles of lovely beaches to walk on—out of season they are usually almost deserted. The mountains are only an hour away, and are higher than any in the British Isles (which is not saying much) and often have snow on them in the winter. The highest ones look down on the desert. In spring, if there has been enough winter rain, the desert bursts into flower. Even at other times it has a strange fascination, partly because of the subtle colors and the wide expanse of sky.

In spite of the almost ideal climate, scientists here seem to work hard. In fact, some of them work so hard that there is no time left for serious thinking. They should heed the saying, "A busy life is a wasted life." I feel much less at home in the rest of America. New York seems almost as remote to me, in both distance and atmosphere, as London now does. My feelings about New York and California are thus just the reverse of Woody Allen's. Woody loves New York and hates California. According to him, "It's only cultural advantage is that you can turn right on a red." But then he seems to enjoy what we in the west call "East Coast tension."

Molecular biology had not stood still in the ten years since 1966, but mostly it had been a period of consolidation. Perhaps the most striking discovery was the retroviruses—RNA viruses that were transcribed onto DNA and incorporated into chromosomal DNA. The key finding was made independently by Howard Temin and David Baltimore. For this they were awarded the Nobel Prize for Medicine in 1975, sharing it with Renato Dulbecco, who is now at the Salk Institute. (The virus that causes AIDS is a retrovirus. Without this pioneer work it would have been difficult to make any sense of AIDS.)

Although I did not appreciate it, molecular biology was on the verge of a massive step forward, caused by three new techniques: recombinant DNA, rapid DNA sequencing, and monoclonal antibodies. Critics who previously had argued that few practical benefits had come from molecular biology were silenced by the realization that, with these new techniques, one could make money out of it. I shall not attempt to describe these very important advances in detail, nor the remarkable results that are now appearing almost every day, mainly because I have not been directly involved in them myself.

I decided that the move to the Salk Institute was an ideal opportunity to become closely interested in the workings of the brain. For many years I had followed parts of the field at a distance. (I first heard of David Hubel and Torsten Wiesel's work on the visual system from a footnote in an article in the literary magazine *Encounter.*) I realized that if I were ever to study the brain more closely it was now or never, since I had just passed sixty.

It took me several years to detach myself from my old interests, especially as in molecular biology surprising things were happening all the time. One of these surprises was the discovery that, in many cases, a stretch of DNA coding for a single polypeptide chain was not continuous, as we had assumed, but was interrupted by long stretches of what appeared to be "nonsense" sequences. These sequences, now called introns, were eliminated from the premessenger RNA by a process called splicing. The resulting messenger RNA, with all the sense bits (called exons) now joined together, was then exported to the cytoplasm so that it could direct the synthesis, on the ribosome, of the protein it coded.

Such introns were found mainly in higher organisms. In our own genes the nonsense sequences (the introns) were often longer than the meaningful ones (the exons). Introns were much sparser in those "higher" organisms, such as the fruit fly *Drosophila,* that had rather little DNA. And in primitive organisms, such as bacteria, introns hardly occurred at all, and then only in special places [small introns in transfer RNA genes].

It was also discovered that not all of the stretches of DNA *between* genes was necessarily very meaningful. Much of our DNA, perhaps as much as 90 percent, appeared at first sight to be unnecessary junk. Even if it had some use, its function probably did not depend on the exact details of its sequence. Leslie Orgel and I wrote an article suggesting that much of it was "Selfish DNA"—a better term might be "Parasitic DNA"—that was there not for the sake of the organism, but for its own sake. Richard Dawkins had already made this suggestion very briefly in a book of his called *The Selfish Gene.*

Leslie and I suggested that this selfish DNA had originated, on many separate occasions, as DNA parasites, which hopped from place to place on the chromosome, leaving replicas of themselves embedded in the host DNA. After a time many of these sequences would be made meaningless by random mutation and then, gradually, over a long period, would be eliminated by the host cell. Meanwhile new parasitic sequences might start to invade the host DNA until eventually a rough balance would be reached between host DNA and parasitic DNA. Whether all this is really true remains to be seen.

The possible existence of such selfish DNA is exactly what might be expected from the theory of natural selection. You are no doubt familiar with the idea of a parasite, such as a tapeworm, but you may not at first accept the idea that a molecule too could be a parasite living in your own chromosomes. But why not?

Notice that the existence of introns came as almost a complete surprise. Nobody had clearly postulated their existence before experimenters stumbled on them by accident. Introns would probably have been discovered earlier if they had existed to any appreciable extent in *E. coli* or in the coli phages. There was no hint of them from classical genetics, even in an organism such as yeast on which relatively high resolution genetic mapping had been carried out. Introns are just the type of thing that is often missed by a pure black-box approach: that is, when only the behavior of the organism is looked at rather than looking inside the organism itself.

During this period I also wrote a scientific book, for lay readers, on the origin of life. Leslie Orgel and I, while attending a scientific meeting on communicating with extraterrestrial intelligence (CETI) held near Yerevan in Soviet Armenia in September 1971, had hit on the idea that perhaps life on Earth originated from microorganisms sent here, on an unmanned spaceship, by a higher civilization elsewhere. Two facts led us to this theory. One was the uniformity of the genetic code, suggesting that at some stage life had evolved through a small population bottleneck. The other was the fact that the age of the universe appears to be rather more than twice the age of the Earth, thus allowing time for life to have evolved twice over from simple beginnings to highly complex intelligence.

We called our theory directed panspermia. Panspermia, a term first used by the Swedish physicist Svante Arrhenius, in 1907, is the idea that microorganisms drifted to the Earth through space and seeded all life on Earth. We used "directed" to imply that someone had deliberately sent the microorganisms here in some way.

The chief difficulty in writing a popular book about the origin of life is that it is mainly a problem in chemistry—mostly organic chemistry. And almost all laymen dislike chemistry. "I understood it all," my mother once said to me about a review I had given her to read, "except for those hieroglyphics." However, the object of my book was not to solve the problem of life's origins but to convey some idea of the many kinds of science involved in the problem, ranging from cosmology and astronomy to biology and chemistry.

I myself had a rather detached view of directed panspermia—I still have—and there was even a passage in the book saying what a good theory should be like and why our theory, though not unprovable, was obviously very speculative. The book, published by Simon & Schuster in 1981, was entitled *Life Itself.* While I considered this title rather too broad for the contents, the publisher insisted on it.

To return to the brain. When I first decided to study it in detail I thought I already knew about most of the problems, at least in broad outline. At Cambridge I had known Horace Barlow for many years—I was introduced to him by my friend Georg Kreisel, the mathematician—and in the fifties had heard Horace talk to the Hardy Club about the frog's eye, with its postulated "insect detectors." At the Hardy Club I had also listened to Alan Hodgkin and Andrew Huxley telling us about their famous model for the axon potential in the squid axon. Later on I met the neurophysiologist David Hubel at a little meeting organized in 1964 at the Salk Institute. The purpose of this meeting was to tell the Salk Fellows what was going on in neurobiology, in case we wanted to make some appointments at the Salk in these fields.

At the same meeting I met for the first time the neurophysiologist Roger Sperry and the neuroanatomist Walle Nauta. There were about a dozen speakers in all, but only a dozen or so listeners, as at that time the Salk was relatively new. However, the listeners were a formidable group, including, for example, Jacques Monod and the physicist Leo Szilard. The audience was so critical that the last speaker was visibly trembling when he took the stand. I only wish that the Salk had been able to start work on neurobiology at that time. While financial considerations made it impossible to do so then, now half its work touches on neurobiology.

I soon found out that what I had learned amounted to very little. Apart from the fact that a lot of work had been done in neuroanatomy and neurophysiology since I had first glanced at these subjects, there were whole areas, such as psychophysics, of which I knew absolutely nothing. (Psychophysics is not some new California religion. It is an old term for that branch of psychology that deals with *measuring* the response of a person or animal to physical inputs, such as light, sound, touch, etc.).

Moreover, I found there was a new subject that called itself cognitive science. (It has been said, somewhat unkindly, that any subject that has "science" in its name is unlikely to be one.) Cognitive science was part of the rebellion against behaviorism. Behaviorists thought that one should study only the behavior of an animal and should not try to take account of, or make models of, any postulated *mental* processes inside the animal. Behaviorism became the dominant school in psychology in the earlier part of this century, especially in America.

Cognitive scientists, in opposition to the narrow views of behaviorists, think it important to make explicit models of mental processes, especially those of humans. Modern linguistics is an important part of cognitive science, since it does just that. There is no great enthusiasm, however, for looking into the actual brain itself. Many cognitive scientists tend to regard the brain as a "black box," better left unopened. In fact, some people *define* cognitive science as studies that take no account of such things as nerve cells. In cognitive science the usual procedure is to isolate some psychological phenomenon, make a theoretical model of the postulated mental processes, and then test the model, by computer simulation, to make sure it works as its author thought it would. If it fits at least some of the psychological facts it is then thought to be a useful model. The fact that it is rather unlikely to be the correct one seems to disturb nobody.

I found all this most peculiar and still do. Basically it is the philosophical attitude of a functionalist, a person who believes that study of the functioning of a person or animal is all important and that it can be studied, by itself, in an abstract way without bothering about what sort of bits and pieces actually implement the functions under study. Such an attitude, I found, is widespread among psychologists. Some even go so far as to deny that knowing exactly what goes on inside the head would ever tell us anything useful at all about psychology. They are apt to bang their fists on the table in support of such statements.

When pressed as to why they think in this way, they usually say that the whole bag of tricks is so fiendishly complicated that no good is likely to come from looking at it closely. The obvious answer to this is that if indeed it is so complicated, how do they ever hope to unscramble the way it operates by looking solely at its input and output, ignoring what goes on between? The only reply I have ever had to such a question is that it is essential to study organisms at higher levels and that the study of neurons, *by itself* (my italics), will never solve such problems. With this I entirely agree, but I cannot see that it justifies ignoring neurons altogether. It is not usually advantageous to have one hand tied behind one's back when tackling a very difficult job.

My own prejudices are exactly the opposite of the functionalists': "If you want to understand function, study structure," I was supposed to have said in my molecular biology days. (I believe I was sailing at the time.) I think one should approach these problems at *all* levels, as was done in molecular biology. Classical genetics is, after all, a black-box subject. The important thing was to combine it with biochemistry. In nature hybrid species are usually sterile, but in science the reverse is often true. Hybrid subjects are often astonishingly fertile, whereas if a scientific discipline remains too pure it usually wilts.

In studying a complicated system, it is true that one cannot even see what the problems are unless one studies the higher levels of the system, but the *proof* of any theory about higher levels usually needs detailed data from lower levels if it is to be established beyond reasonable doubt. Moreover, exploratory data from the study of lower levels often suggests important ways of constructing new higher-level theories. In addition, useful information about lower-level components can often be obtained from studying them in simpler animals, which may be easier to work on. An example would be recent work on the mechanism of memory in invertebrates.

My first problem was to decide what sort of animal to concentrate on. Some of my fellow molecular biologists had opted for small, rather primitive animals. As mentioned, Sydney Brenner had selected a nematode. Seymour Benzer had chosen to study the behavioral genetics of the little fruit fly, *Drosophila,* partly because so much basic genetics had already been done on it.

I decided that my main *long-term* interest was in the problem of consciousness, though I realized that it would be foolish to start with this. Consciousness is most apparent in man—at least I know I am conscious and I have good reasons to suspect that you are too. Whether a fruit fly is conscious is an open question. There are, however, grave experimental handicaps to working on human beings, since so many experiments are impossible for ethical reasons. It seemed reasonable, therefore, to concentrate on animals close to man in evolution; that is, the mammals and in particular the primates—the monkeys and apes.

My next problem was to choose some particular aspect of the mammalian brain. As I knew very little I decided to make the obvious choice and concentrate on the visual system. Man is a very visual animal (as are monkeys), and much work had already been done on many aspects of vision.

How can one study vision in man by working on monkeys? The obvious approach is to do what one can on man, and, in parallel, study the same system in a monkey or other mammal. In work on perception, it is now becoming standard practice to use arguments from detailed psychophysical studies on man (plus rather cruder psychophysical studies on a monkey) combined with all the neuroanatomical and neurophysiological knowledge available on the relevant part of a monkey's brain. Occasionally other data from man can be used, such as evoked potentials (a type of brain wave), or various rather expensive scans, but as yet these have a much lower resolution, either in space and/or time, and thus usually give us much less information.

This is why, to someone like myself, the visual system is attractive since, as far as we can tell, a macaque monkey sees in much the same way as we do. There are, of course, few subjects more important to us than language, since it is one of the main differences between man and all lower animals. Unfortunately, for this very reason, there is no suitable animal for such studies. This is why I believe that modern linguistics, sophisticated though it is, will run into a brick wall unless much more can be found out about what happens inside our heads when we talk, listen to speech, and read. If language is anything like as complex as vision (which seems more than likely), the chance of unscrambling the way it really works without this extra knowledge seems to me to be rather small. Linguists, not surprisingly, usually find this argument unacceptable.

I also decided that, at least at first, I would not attempt to do experiments. Apart from the fact that, technically, they are often very difficult, I thought I could contribute more from a theoretical viewpoint. It seemed to me that I might perform a useful function by studying the problem of vision from as many points of view as possible. I hoped that I might help to build bridges between the various scientific disciplines, all of which studied the brain from one point of view or another. I had rather little expectation of producing any radically new theoretical ideas at such an advanced age, but I thought I might interact fruitfully with younger scientists. In any case I expected that the subject would prove endlessly interesting and that at my time of life I had a right to do things for my own amusement, provided I could make an occasional useful contribution.

Having decided that I could learn about the mammalian visual system, my next problem was to select which aspect to study first. I had never had a medical education, so my knowledge of neuroanatomy was almost zero. I decided to tackle that first, as I expected it to be the dullest part of the subject. It would be as well, I thought, to get it out of the way before going on to other, more interesting, topics.

To my surprise I soon discovered that there had been a minor revolution in dry-as-dust neuroanatomy. Thanks mainly to the introduction of various rather simple biochemical techniques, it was now possible to discover how the various regions of the brain were connected together. Moreover, the techniques were not only powerful but considerably more reliable than most of the older methods. Unfortunately most of them cannot be used on humans (one cannot, at the end of an experiment, "sacrifice" the graduate student who has been acting as the subject, as one can do with animals, for obvious ethical reasons). We thus have the curious situation that more is known about neural connections in the brain of the macaque monkey than about those in the human brain. In fact, we shall soon know so much about the broad pattern of connections in the macaque, and about the location in the brain of various chemical transmitters and the receptors for them, that the only way to cope with all this new information will be to store it in computers, in such a way that it can be displayed in some vivid graphic form for easy comprehension.

I first started by reading experimental papers and reviews. I found it was not difficult to approach experimentalists provided one was genuinely interested in what they were doing and had first made some effort to discover from their publications what they were up to. In this way I made many new friends, far too many to list here. I was lucky in finding in La Jolla several people interested in vision or in theory. A group at the Psychology Department at the University of California, San Diego (UCSD), under the leadership of Bob Boyton, studied mainly the psychophysics of vision. Other psychophysicists I got to know were Don MacLeod and V. S. ("Rama") Ramachandran when he came to San Diego from Irvine. I also interacted with another group in the same department, led then by David Rumelhart and Jay McClelland, that did theoretical work. After a while the department appointed me an adjunct professor of psychology, in spite of my very flimsy knowledge of the subject.

In 1980 Max Cowan came to the Salk, setting up a large group of neuroscientists there. Some of these people, such as Richard Andersen (now at M.I.T.) and Simon LeVay do experimental work on the visual system. Although Max left in 1986, the Salk still has a strong interest in neuroscience and has recently recruited Tom Albright, an experimentalist from Princeton.

Another blessing was the arrival, in 1984, of the Canadian philosophers Paul and Pat Churchland, to take up chairs in the philosophy department at UCSD. It is unusual to find philosophers who are even remotely concerned about the brain, so it is a great help to have the advice of two people who do take a keen interest in it. Both had written very well on reductionism (a dirty word to some, especially to those who regard me as an archreductionist). More recently Pat has written a large book, called *Neurophilosophy*, published by the Bradford Book section of the M.I.T. Press, setting out the philosophical, theoretical, and experimental aspects of their new point of view. Its subtitle is "Towards a Unified Science of the Mind-Brain."

Ramachandran and Gordon Shaw (a physicist at U.C. Irvine) were the co-founders of the Helmholtz Club, named after the nineteenth-century German physicist who pioneered the scientific study of perception. The members meet about once a month, starting with lunch and ending with dinner. In between we have talks by two speakers, on topics mostly connected with the visual system. This schedule allows plenty of time for discussion. The meetings are held at Irvine, which is midway between Los Angeles and San Diego, so that members and guests from the other university campuses can attend.

This is not the place for me to attempt to outline what we now know about the visual system—that would take at least another whole book—let alone the rest of the brain. I will restrict myself to rather general comments. In the first place, it is not obvious to most people why we need to study vision. Since we see so clearly, without any apparent effort, what is the problem? It comes as a surprise to learn that in order to construct our vivid mental representation of the outside world, the brain has to engage in many complex activities (sometimes called computations) of which one is almost completely unaware.

We succumb all too easily to the Fallacy of the Homunculus—that somewhere attached to our brain there is a little man who is watching everything that is going on. Most neuroscientists don't believe this (Sir John Eccles is an exception) and think that our picture of the world and of ourselves is due solely to neurons firing and other chemical or electrochemical processes inside one's body. Exactly how these activities give us our vivid picture of the world and of ourselves and also allow us to act is what we want to discover.

The main function of the visual system is to build a representation inside our head of objects in the world outside us. It has to do this from the complex signals reaching the retinas of our eyes. Though these signals have much information implicit in them, the brain needs to process this information to obtain *explicit* representations of what interests it. Thus the photoreceptors in our eyes respond to the wavelength of the impinging light coming from an object. But what the brain is mainly interested in is the *reflectivity* (the color) of an object, and it can extract this information even under quite different conditions of illumination of that object.

The visual system has been evolved to detect those many aspects of the real world that, in evolution, have been important for survival, such as the recognition of food, predators, and possible mates. It is especially interested in moving objects. Evolution will latch onto any features that will give useful information. In many cases the brain has to perform its operations as quickly as possible. The neurons themselves are inherently rather slow (compared to transistors in a digital computer) and so the brain has to be organized to carry out many of its "computations" as quickly as possible. Exactly how it does this we do not yet understand.

It is very easy to convince someone that however he may think his brain works, it certainly doesn't work like that. That misunderstanding can be demonstrated from the effects of human brain damage, or by psychophysical experiments on undamaged humans, or by outlining what we know about monkey brains. What seems a uniform and simple process is in fact the result of elaborate interactions between systems, subsystems, and sub-subsystems. For example, one system determines how we see color, another how we see in three dimensions, (although we receive only two-dimensional information from each of our two eyes), and so on. One of the subsystems of the latter depends on the *difference* between the images in our two eyes; this is called stereopsis. Another deals with perspective. Another uses the fact that objects at a distance subtend a smaller angle than when they are nearer to us. Others deal with occlusion (one object occluding part of an object behind it), shape-from-shading, and so on. Each of these subsystems may well need sub-subsystems to make it work.

Normally all the systems produce roughly the same answer, but by using tricks, such as constructing rather artificial visual scenes, we can pit them against one another and so produce a visual illusion. If a person looks, with one eye through a small hole, into a room built with false perspectives, an object on one side of the room can be made to appear smaller than the same object on the other side. Such a full-scale room, called an Ames room, exists at the Exploratorium in San Francisco. When I was looking into it some children appeared to be running from side to side. They appeared to grow taller as they ran to one side and to get shorter again as they ran back to the other side. Of course, I know full well that children never change height in this way, but the illusion was nevertheless completely compelling.

The conception of the visual system as a bag of tricks has been put forward by Rama Ramachandran, mainly as a result of his elegant and ingenious psychophysical studies. He calls his point of view the utilitarian theory of perception, writing:

> It may not be too farfetched to suggest that the visual system uses a bewildering array of special-purpose tailor-made tricks and rules-of-thumb to solve its problems. If this pessimistic view of perception is correct, then the task of vision researchers ought to be to uncover these rules rather than to attribute to the system a degree of sophistication that it simply doesn't possess. Seeking overarching principles may be an exercise in futility.

This approach is at least compatible with what we know of the organization of the cortex in monkeys and with François Jacob's idea that evolution is a tinkerer. It is, of course, possible that underlying all the various tricks there are just a few basic learning algorithms that, building on the crude structures produced by genetics, produce this complicated variety of mechanisms.

Another thing I discovered was that although much is known about the behavior of neurons in many parts of the visual system (at least in monkeys), nobody really has any clear idea how we actually see anything at all. This unhappy state of affairs is usually never mentioned to students of the subject. Neurophysiologists have some glimpses into how the brain takes the picture apart, how somewhat separate areas of our cerebral cortex process motion, color, shape, position in space, and so on. What is not yet understood is how the brain puts all this together to give us our vivid unitary picture of the world.

I also discovered that there was another aspect of the subject one was not supposed to mention. This was consciousness. Indeed an interest in the topic was usually taken as a sign of approaching senility. This taboo surprised me very much. Of course, I knew that until recently most of the experiments on the visual system of animals were done when the animals were unconscious under an anesthetic so that, strictly speaking, they could not see anything at all. For many years this did not unduly disturb the experimentalists, since they found that the neurons in the brain, even under these restrictive conditions, behaved in such interesting ways. Recently more work has been done on alert animals. Although these animals are technically rather more difficult to study, there are compensations, since the animals are returned to their cages after a normal day's work and the experimenter can go home to supper. Such animals are usually studied for many months before being sacrificed. (Experiments on anesthetized animals can be much more demanding since they usually last for many many hours at a stretch, after which the animal is sacrificed straight away.) Curiously enough, hardly any experiments have yet been done on the *same* sort of neurons, in the *same* animal, first when it is alert and then when it is under an anesthetic.

It was not only neurophysiologists who disliked talking about consciousness. The same was true of psychophysicists and cognitive scientists. A year or so ago the psychologist George Mandell did organize a course of seminars at the psychology department at UCSD. The seminars showed that there was hardly any consensus as to what the problem was, let alone how to solve it. Most of the speakers seemed to think that no solution was possible in the near future and merely talked around the subject. Only David Zipser (another ex-molecular biologist, now at UCSD) thought as I did, namely that consciousness was likely to involve a special neural mechanism of some sort, probably distributed over the hippocampus and over many areas of the cortex, and that it was not impossible to discover by experiment at least the general nature of the mechanism.

Curiously enough, in biology it is sometimes those basic problems that look impossibly difficult to solve which yield the most easily. This is because there may be so few even remotely possible solutions that eventually one is led inexorably to the correct answer. (An example of such a problem is discussed toward the end of chapter 3.) The biological problems that are really difficult to unscramble are those where there is almost an infinity of plausible answers and one has painstakingly to attempt to distinguish between them.

One main handicap to the experimental study of consciousness is that while people can tell us what they are conscious of (whether they have suddenly lost their color vision, for example, and now only see everything in shades of gray), it is more difficult to obtain this information from monkeys. True, monkeys *can* be laboriously trained to press one key if they see a vertical line and another if they are shown a horizontal one. But we can ask people to *imagine* color, or to imagine they are waggling their fingers. It is difficult to instruct monkeys to do this. And yet we can look inside a monkey's head in much more detail than we can look inside a person's head. It is therefore not unimportant to have some *theory* of consciousness, however tentative, to guide experiments on both humans and monkeys. I suspect that consciousness may be able to do without a fully working long-term memory system but that very short-term memory is indispensable to it. This suggests straight away that one should look into the molecular and cellular basis of very short-term memory—a rather neglected subject—and this can be done on animals, even on a cheap and relatively simple animal like a mouse.

And what of theory? It is easy to see that theory of some sort is essential, since any explanation of the brain is going to involve large numbers of neurons interacting in complicated ways. Moreover, the system is highly nonlinear, and it is not easy to guess exactly how any complex model will behave.

I soon found that much theoretical work was going on. It tended to fall into a number of somewhat separate schools, each of which was rather reluctant to quote the work of the others. This is usually characteristic of a subject that is not producing any definite conclusions. (Philosophy and theology might be good examples.) I renewed acquaintance with the theorist David Marr (whom I had originally met in Cambridge) when he came with another theorist, Tomaso (Tommy) Poggio, to the Salk for a month in April 1979 to talk about the visual system. Alas, David is now dead, at the early age of thirty-five, but Tommy (now at M.I.T.) is still alive and well, and has become a close friend. Eventually I met many of the theorists working on the brain (too numerous to list here), mainly by going to meetings. Some I got to know better from personal visits.

Much of this theoretical work was on neural nets—that is, on models in which groups of units (somewhat like neurons) interact in complicated ways to perform some function connected, often rather remotely, with some aspect of psychology. Much work was being done on how such nets could be made to learn, using simple rules—algorithms—devised by the theorists.

A recent two-volume book, entitled *Parallel Distributed Processing* (PDP), describes much of the work done by one school of theorists, the San Diego group and their friends. It is edited by David Rumelhart (now at Stanford) and Jay McClelland (now at Carnegie-Mellon) and published by Bradford Books. For such a large, rather academic book it has proved to be a best-seller. So striking are the results that the PDP approach is having a dramatic impact both on psychologists and on workers in artificial intelligence (AI), especially those trying to produce a new generation of highly parallel computers. It seems likely to become the new wave in psychology.

There is no doubt that very suggestive results have been produced. For example, we can see how a neural net can store a "memory" of various firing patterns of its "neurons" and how any small part of one of the patterns (the cue) can recall the entire pattern. Also how such a system can be taught by experience to learn tacit rules (just as a child first learns the rules of English grammar tacitly, without being able to state them explicitly). One example of such a net, called NetTalk, set up by Terry Sejnowski and Charles Rosenberg, gives rather a striking demonstration of how this little machine can, by experience, learn to pronounce correctly a written English text, even one it has never seen before. Terry, whom I got to know well, gave a striking demonstration of it one day at a Salk Faculty lunch. (He has also talked about it on the *Today* show.) This simple model doesn't *understand* what it is reading. Its pronunciation is never completely correct, partly because, in English, pronunciation sometimes depends on meaning.

In spite of this I have some strong reservations about the work done so far. In the first place, the "units" used almost always have some properties that appear unrealistic. For example, a single unit can produce excitation at some of its terminals and inhibition at others. Our present knowledge of the brain, admittedly limited, suggests that this seldom if ever happens, at least in the neocortex. It is thus impossible to test all such theories at the neurobiological level since at the very first and most obvious test they fail completely. To this the theorists usually reply that they could easily alter their models to make that aspect of them more realistic, but in practice they never bother to do this. One feels that they don't really want to know whether their model is right or not. Moreover, the most powerful algorithm now being used [the so-called back-propagation algorithm] also looks highly unlikely in neurobiological terms. All attempts to overcome this particular difficulty appear to me to be very forced. Reluctantly I must conclude that these models are not really theories but rather are "demonstrations." They are existence proofs that units somewhat like neurons can indeed do surprising things, but there is hardly anything to suggest that the brain actually performs in exactly the way they suggest.

Of course, it is quite possible that these nets and their algorithms could be used in the design of a new generation of highly parallel computers. The main technical problem here seems to be to find some neat way to embody *modifiable* connections in silicon chips, but this problem will probably be solved before long.

There are two other criticisms of many of these neural net models. The first is that they don't act fast enough. Speed is a crucial requirement for animals like ourselves. Most theorists have yet to give speed the weight it deserves. The second concerns relationships. An example might help here. Imagine that two letters—*any* two letters—are briefly flashed on a screen, one above the other. The task is to say which one is the upper one. (This problem has been suggested independently by the psychologists Stuart Sutherland and Jerry Fodor.) This is easily done by older models, using the processes commonly employed in modern digital computers, but attempts to do it with parallel distributed processing appear to me to be very cumbersome. I suspect that what is missing may be a mechanism of *attention.* Attention is likely to be a *serial* process working on top of the highly parallel PDP processes.

Part of the trouble with theoretical neuroscience is that it lies somewhat between three other fields. At one extreme we have those researchers working directly on the brain. This is science. It is attempting to discover what devices nature actually uses. At the other extreme lies artificial intelligence. This is engineering. Its object is to *produce a device* that works in the desired way. The third field is mathematics. Mathematics cares neither for science nor for engineering (except as a source of problems) but only about the relationship between abstract entities.

Workers in brain theory are thus pulled in several directions. Intellectual snobbery makes them feel they should produce results that are mathematically both deep and powerful and also apply to the brain. This is not likely to happen if the brain is really a complicated combination of rather simple tricks evolved by natural selection. If an idea they conceive doesn't help to explain the brain, the theorists may hope that perhaps it may be useful in AI. There is thus no compelling drive for them to press on and on until the way the brain actually works is laid bare. It is more fun to produce "interesting" computer programs and much easier to get grants for such work. There is even the possibility that they might make some money if their ideas could be used in computers. The situation is not helped by the general view that psychology is a "soft" science, which seldom if ever produces definitive results but stumbles from one theoretical fad to the next one. Nobody likes to ask if a model is really correct since, if they did, most work would come to a halt.

I wish I could say that my own efforts amounted to much. From thinking about neural nets Graeme Mitchison and I invented in 1983 a new reason for the existence of rapid-eye-movement (REM) sleep, though two other groups independently thought of the same mechanism. This is a lot of fun to lecture about since almost everybody is interested in sleep and dreams. I have given the lecture to physicists (including the research department of an oil company), women's clubs, and high-school teachers as well as to numerous academic departments. The essence of the idea is that memories are likely to be stored in the mammalian brain in a very different way from the way they are stored in a filing system or in a modern computer. It is widely believed that, in the brain, memories are both "distributed" and to some extent superimposed. Simulations show that this need not cause a problem unless the system becomes overloaded, in which case it can throw up false memories. Often these are mixtures of stored memories that have something in common.

Such mixtures immediately remind me of dreams and of what Freud called condensation. For example, when we dream of someone, the person in the dream is usually a mixture of two or three rather similar people. Graeme and I therefore proposed that in REM sleep (sometimes called dream sleep), there is an automatic correction mechanism that acts to reduce this possible confusion of memories. We suggest that this mechanism is the root cause of our dreams, most of which, incidentally, are not remembered at all. Whether this idea is true or not only time will tell.

I also wrote a paper on the neural basis of attention, but this also is highly speculative. I have yet to produce any theory that is both novel and also explains many disconnected experimental facts in a convincing way.

Looking back, I can recall how very strange I found this new field. There is no doubt that, compared to molecular biology, brain science is in an intellectually backward state. Also the pace is much slower. One can see this by noting the use of the word recently. In classical studies (Latin and Greek) "recently" means within the last twenty years. In neurobiology or psychology it usually means within the last few years, whereas in modern molecular biology it means within the last few *weeks*.

Three main approaches are needed to unscramble a complicated system. One can take it apart and characterize all the isolated bits—what they are made of and how they work. Then one can find exactly where each part is located in the system in relation to all the other parts and how they interact with each other. These two approaches are unlikely, by themselves, to reveal exactly how the system works. To do this one must also study the behavior of the system and its components while interfering very delicately with its various parts, to see what effect such alterations have on behavior at all levels. If we could do all this to our own brains we would find out how they work in no time at all.

Molecular and cell biology could help decisively in all these three approaches. The first has already begun. For example, the genes for a number of the key molecules have already been isolated, characterized, and their products produced so that they can be more easily studied. A little progress has been made on the second approach, but more is still needed. For example, a technique for injecting a single neuron in such a way that all the neurons connected to it (and only those) are labeled would be useful.

The third approach also needs new methods, especially as the usual ways of ablating parts of the brain are so crude. For example, it would be useful to be able to inactivate, preferably reversibly, a single type of neuron in a single area of the brain. In addition, more subtle and powerful ways of studying behavior, both of the whole animal and also of groups of neurons, are needed. Molecular biology is advancing so rapidly that it will soon have a massive impact on all aspects of neurobiology.

In the summer of 1984 I was asked to address the Seventh European Conference on Visual Perception in Cambridge, England. It was one of those after-dinner occasions when one is expected to entertain as well as to inform. I finished by stating that in a generation's time most of the researchers in psychology departments would be working on "molecular psychology." I could see expressions of total disbelief on the faces of most of my audience. "If you don't accept that," I said, "look what has happened to *biology* departments. Nowadays most of the scientists there are doing *molecular* biology, whereas a generation ago that was a subject known only to specialists." Their disbelief changed to apprehension. Is *that* what the future had in store? The last couple of years has shown that the beginning of this trend is already with us [recent work on the NMDA receptor for glutamate and its relation to memory, for example].

The present state of the brain sciences reminds me of the state of molecular biology and embryology in, say, the 1920s and 1930s. Many interesting things have been discovered, each year steady progress is made on many fronts, but the major questions are still largely unanswered and are unlikely to be without new techniques and new ideas. Molecular biology became mature in the 1960s, whereas embryology is only just starting to become a well-developed field. The brain sciences have still a very long way to go, but the fascination of the subject and the importance of the answers will inevitably carry it forward. It is essential to understand our brains in some detail if we are to assess correctly our place in this vast and complicated universe we see all around us.

APPENDIX A

A Brief Outline of Classical Molecular Biology

THE GENETIC MATERIAL of all organisms in nature is nucleic acid. There are two types of nucleic acid: DNA (short for deoxyribonucleic acid) and the closely related RNA (short for ribonucleic acid). Some small viruses use RNA for their genes. All other organisms and viruses use DNA. (Slow viruses may be an exception.)

The molecules of both DNA and RNA are long and thin, sometimes extremely long. DNA is a polymer, with a regular backbone, having alternating phosphate groups and sugar groups (the sugar is called deoxyribose).

To each sugar group is attached a small, flat, molecular group called a base. There are four major types of base, called A (adenine), G (guanine), T (thymine), and C (cytosine). (A and G are purines; T and C are pyrimidines.) The *order* of the bases along any particular stretch of DNA conveys the genetic information. By 1950 Erwin Chargaff had discovered that in DNA from many different sources the amount of A equaled the amount of T, and the amount of G equaled the amount of C. These regularities are known as Chargaff's rules.

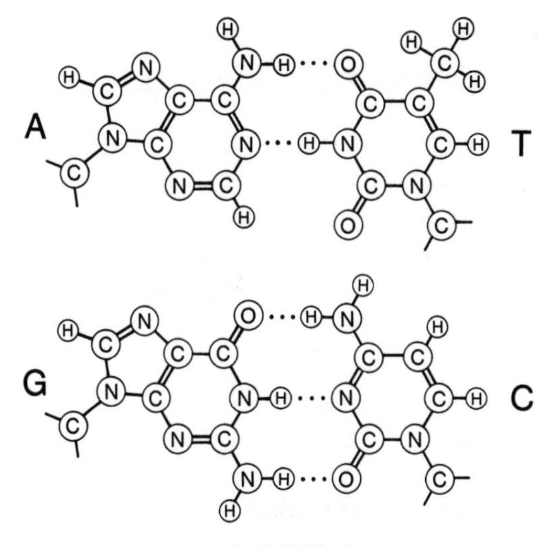

FIGURE A–1

The two base pairs: A=T and G≡C. For the bases: A—Adenine, T—Thymine, G—Guanine, C—Cytosine. For the atoms: C—Carbon, N—Nitrogen, O—Oxygen, H—Hydrogen.

RNA is similar in structure to DNA, except that the sugar is slightly different (ribose instead of deoxyribose) and instead of T there is U (uracil). (Thymine itself is 5-methyl-uracil.) Thus, the AT pair is replaced by the very similar AU pair.

DNA is usually found in the form of a double helix, having two distinct chains wound around one another about a common axis. Surprisingly, the two chains run in *opposite* directions. That is, if the sequence of atoms in the backbone of one chain runs up, then that of the other runs down.

At any level the bases are paired. That is, a base on one chain is paired with the base opposite it on the other chain. Only certain pairs are possible. They are:

$$\left\{\begin{array}{ccc} T & = & A \\ A & = & T \end{array}\right.$$

$$\left\{\begin{array}{ccc} G & \equiv & C \\ C & \equiv & G \end{array}\right.$$

Their chemical formulas are shown in figure A–1. These base pairs are held together by weak bonds, called hydrogen bonds, symbolized here by the dashes. Thus the AT pair forms two hydrogen bonds, the GC pair three of them. *This pairing of the bases is the key feature of the structure.*

To replicate DNA, the cell unwinds the chains and uses each single chain as a template to guide the formation of a new companion chain. After this process is completed we are left with *two* double helices, each containing one old chain and one new one. Since the bases for the new chains must be selected to obey the pairing rules (A with T, G with C), we end up with *two* double helices, each identical in base sequence to the one we started with. In short, this neat pairing mechanism is the molecular basis for like reproducing like. The actual process is a lot more complicated than the outline just sketched.

A major function of nucleic acid is to code for protein. A protein molecule is also a polymer, with a regular backbone (called a polypeptide chain) and side groups attached at regular intervals. Both backbone and side-chains of protein are quite different chemically from the backbone and side groups of nucleic acid. Moreover, there are twenty different side groups found in proteins, compared with only four in nucleic acid.

The general chemical formula of a polypeptide chain is shown in figure A–2. The "side-chains" are attached at the points marked R, R′, R″, and so forth. The exact chemical formula of each of the twenty different side-chains is known and can be found in any textbook of biochemistry.

Each polypeptide chain is formed by joining together, head to tail, little molecules called amino acids. The general formula for an amino acid appears in figure A–3, where R represents the side-chain that is different for each of the magic twenty. During this process one molecule of water is eliminated as each join is made. (The actual chemical steps are a little more complicated than this simple, overall description.)

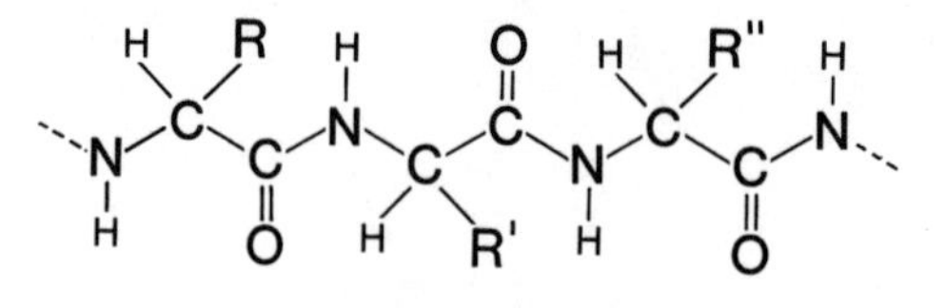

FIGURE A–2

The basic chemical formula for a polypeptide chain (approximately three repeats are shown). C—Carbon, N—Nitrogen, O—Oxygen, H—Hydrogen. R, R′, R″—the various side-chains (R stands for Residue).

All the amino acids inserted into proteins (except glycine) are L-amino acids, as opposed to their mirror images, which are called D-amino acids. This terminology refers to the three-dimensional configuration around the upper carbon atom in figure A–3.

The synthesis of a protein takes place on a complicated piece of biochemical machinery, called a ribosome, aided by a set of small RNA molecules called tRNA (transfer RNA) and a number of special enzymes. The sequence information is provided by a type of RNA molecule called mRNA (messenger RNA). In most cases this mRNA, which is single-stranded, is synthesized as a copy of a particular stretch of DNA, using the base-pairing rules. A ribosome travels along a piece of mRNA, reading off its base sequence in groups of three at a time, as explained in appendix B. The overall process is DNA ~~> mRNA ~~> protein, where the wiggly arrows show the direction in which the sequence information is transferred.

To make matters even more complicated, each ribosome is constructed not only with a large set of protein molecules but also with several molecules of RNA, two of them being fairly large. These RNA molecules are *not* messengers. They form part of the ribosomal structure.

As a polypeptide chain is synthesized, it folds itself up to form the intricate three-dimensional structure that protein needs to perform its highly specific function.

Proteins come in all sizes. A typical one might be several hundred side groups long. Thus a gene is often a stretch of DNA, typically about a thousand or more base pairs long, that codes for a single polypeptide chain. Other parts of the DNA are used as control sequences, to help turn particular genes on and off.

The nucleic acid of a small virus may be about 5,000 bases long,

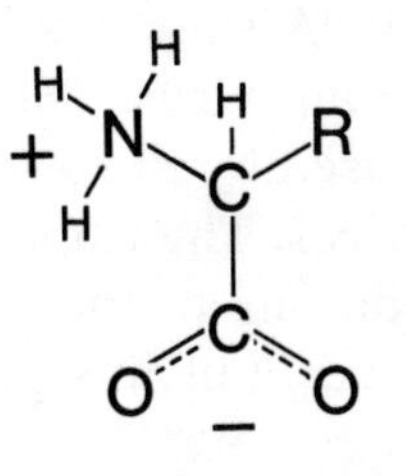

FIGURE A–3

The general formula for an amino acid. The amino group is NH_3^+. The acid group is COO^-. The side-group, which differs from one amino acid to another, is denoted R. C—Carbon, N—Nitrogen, O—Oxygen, H—Hydrogen.

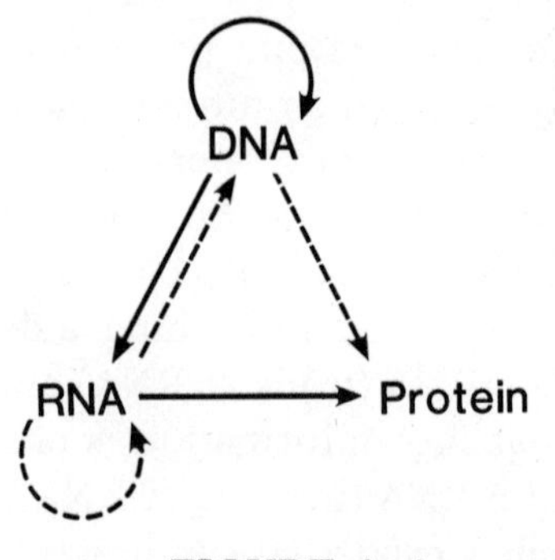

FIGURE A–4

A diagram to illustrate the central dogma. The arrows represent the various transfers of sequence information. Solid arrows show the common transfers. Dotted arrows show the rarer transfers. Notice that the missing transfers are those for which the arrows would start from protein.

and will code for a handful of proteins. A bacterial cell is likely to have some million bases in its DNA, often all in one circular piece, and code for several thousand different kinds of protein. One of your own cells has about three billion bases from your mother and a similar number from your father, coding for some 100,000 kinds of proteins. It was discovered in the 1970s that the DNA of higher organisms may contain long stretches of DNA (some of which occur *within* genes, and are called introns) with no apparent function.

The so-called central dogma is a grand hypothesis that attempts to predict which transfers of sequence information *cannot* take place. These correspond to the missing arrows in figure A–4. The common transfers are shown by the solid lines; the rarer ones by the dotted ones. Note that the missing arrows correspond to all possible transfers *from* protein.

The common transfers have been described earlier. Of the rarer ones the transfer RNA ~> RNA is used by certain RNA viruses, such as the flu virus and the polio virus. The transfer RNA ~> DNA (reverse transcription) is used by the so-called RNA retroviruses. An example is the AIDS virus. The transfer DNA ~> protein is a freak. Under special conditions in the test tube, single-stranded DNA can act as a messenger, but this probably never occurs in nature.

APPENDIX B

The Genetic Code

THE GENETIC CODE is the little dictionary that relates the four-letter language of the nucleic acids (A, G, T, and C for DNA; RNA has U in place of T) to the twenty-letter language of proteins. A group of three adjacent letters, called a codon, codes for an amino acid. (There are $4 \times 4 \times 4 = 64$ codons in all.) Most amino acids are coded by more than one codon. In addition, three codons stand for "end of chain."

The genetic code is usually displayed as shown in table B–1. At first sight the table may appear rather confusing, but basically it is very simple. The exact chemical formula of each amino acid is known. For example, one of the amino acids is called valine. To make the table easier to read, valine is abbreviated to Val. In a similar way histidine, the name of one of the other amino acids, is written His. The three bases of each triplet can be read off for each entry in the table. The first base is written on the left, the second at the top, and the third on the right. Thus it can be seen that valine (Val) is coded by GUU, GUC, GUA, and GUG, whereas histidine (His) has the codons CAU and CAC. The three codons for ending the polypeptide chain (STOP) are UAA, UAG, and UGA. The left end of an RNA or DNA chain, as usually written, is called the 5′ end and the right end the 3′ end, for chemical reasons.

1st position (5′ end) ↓	2nd position U	C	A	G	3rd position (3′ end) ↓
U	Phe	Ser	Tyr	Cys	U
	Phe	Ser	Tyr	Cys	C
	Leu	Ser	STOP	STOP	A
	Leu	Ser	STOP	Trp	G
C	Leu	Pro	His	Arg	U
	Leu	Pro	His	Arg	C
	Leu	Pro	Gln	Arg	A
	Leu	Pro	Gln	Arg	G
A	Ile	Thr	Asn	Ser	U
	Ile	Thr	Asn	Ser	C
	Ile	Thr	Lys	Arg	A
	Met	Thr	Lys	Arg	G
G	Val	Ala	Asp	Gly	U
	Val	Ala	Asp	Gly	C
	Val	Ala	Glu	Gly	A
	Val	Ala	Glu	Gly	G

TABLE B–1

The code appears to be exactly the same for all higher plants and animals studied so far. Nevertheless, minor variations are known, especially for the DNA of certain mitochondria, the little organelles that live in the cytoplasm of higher organisms and for certain of the fungi.

Abbreviations

U Uracil (for DNA, read T [=Thymine] instead of U)
C Cytosine
A Adenine
G Guanine

Ala Alanine
Arg Arginine
Asn Asparagine
Asp Aspartic acid
Cys Cysteine
Gln Glutamine
Glu Glutamic acid
Gly Glycine
His Histidine
Ile Isoleucine
Leu Leucine

Lys Lysine
Met Methionine
Phe Phenylalanine
Pro Proline
Ser Serine
Thr Threonine
Trp Tryptophan
Tyr Tyrosine
Val Valine

STOP means "end chain"

Index

Acridines, 125, 135–36
Adaptor hypothesis, 95–98
Adenine, 62, 65, 90, 93, 97, 99–100, 128–29, 132, 134, 164–66, 169–70
Admiralty, 14–17, 19, 36, 43
Adrian, Lord, 18
Alanine, 54, 91, 113, 170
Albright, Thomas, 153
Algorithms, 159–60
Ames room, 156
Amino acids, 18, 34, 167; adaptors and, 96; alteration of single, 103, 105; DNA and RNA templates for side-chains of, 96; head-tail joining of in polypeptide chain, 54; mutants and, 126; one-dimensional sequences of, 36; polypeptide chains and, 166; in proteins, 170; sequences of, 36, 89, 93, 97, 102–3, 108; side-chains, 95; total number of, 89, 91–92, 100, 130; tRNA and, 113; *see also specific amino acids*
Amylase, 32
Andersen, Richard, 153
Andrade, Edward Neville da Costa, 13
Antibiotics, advances in, 17
Antibodies, monoclonal, 146
Arginine, 54, 170
Arrhenius, Svante, 148
Artificial intelligence (AI), 159–61
Asimov, Isaac, 82
Asparagine, 91, 170
Astbury, William, 53–55, 64; α helix and, 57
Astrachan, Lazarus, 119
Astronomy, 137, 148
Asymmetric unit, 65
Atomic energy, 19
Atoms: Bohr theory, 13; carbon, 61; distance between, 20; energies between, 140; exact configuration of in space, 55; heavy, 49–50, 103; hydrogen, 54, 65; light, 50; mercury, 49–50; nitrogen, 59, smashing of by Cockcroft and Walton, 39
Attention, neural basis of, 162
Autocorrelation function, of electron density, 48
Avery, Oswald, 36–37

Bachmann, Larry, 82–84
Backmutation, 132
Back-propogation algorithm, 160
Bacteria, 117, 168; blender experiment and, 118; diaminopimelic acid found in cell walls of, 91; difference between genotype and phenotype and, 107; gene control and, 143; genetics and, 36–37, 64, 118; introns and, 147; lysis of, 104; rough coat, 37; smooth coat, 36–37; *see also* Bacteriophages; Phages
Bacterial Cell, The (Hinshelwood), 18
Bacteriophages, 61; T4, 119, 122–24; *see also* Bacteria; Phages
Baltimore, David, 146
Barlow, Horace, 148
Barnett, Leslie, 133–35
Base pairs: DNA, 27, 62, 65–66, 71, 75, 111–12, 122, 165–67; mutation rate in effective, 27; pairing rules for, 63, 75, 165–67; replication errors and, 111; RNA, 27; *see also* Nucleotides
Bauer, William, 73
Beadle, George, 33
Behavioral genetics, 151

Behaviorism, 149
Benzer, Seymour, 98, 106, 123–24, 126; behavioral genetics of *Drosophila* and, 151; Biochemical Congress at Moscow (1961) and, 131; one-dimensionality of gene and, 122, 124
Bernal, J. D., 21, 40–41, 48–49, 51, 75
Bessel functions, 67
β-galactosidase, 118, 120
Big Bang theory, 94
Biochemical Congress, Moscow (1961), 130–31
Biochemistry, 21, 34, 37, 63, 75, 77, 93, 135
Biological replication, natural selection and, 5
Biology: development of theory in, 139; difference of from physics, 5–6, 137–39; entry of physicists into, 19; hydrogen bond and, 18; laws of, 5, 138; mechanisms of, 138
Biophysics, 13, 19–21
Birkbeck College, London, 21
Black box: brain as, 149; classical genetics as, 150
Blender experiment, 118
Blending inheritance, 26–27
Blind Watchmaker, The (Dawkins), 28–29
Blue-green algae, 144; *see also* Bacteria
Bohr, Niels, 13
Bondi, Hermann, 94
Bonds: atomic, 61; distances of, 55; double, 55; electrostatic interactions of, 61; homopolar, 61; hydrogen, 56, 61–62, 71, 95, 115, 165; multiple, 56; peptide, 58; single, 55; and van der Waal's forces, 61
Botany, 9
Boyton, Robert, 153
Brachet, Jean, 140
Bragg, Sir Lawrence, 23, 40–41, 47, 50–51, 53, 55, 63, 77, 85; α helix and, 53–54, 57–58
Bragg, Sir William, 40
Bragg's law for X-ray diffraction, 40, 46
Brain: ablation of, 162; as black box, 149; cerebral cortex, 156, 157; hippocampus, 157; mammalian, 150; memory and, 161; models of, 115; neocortex, 159; as product of natural selection, 161; receptors, 153; reflectivity and, 155; sciences, 163; transmitters, 153; visual system, 150–58; waves, 150; workings of, 17–18, 146, 148–49, 158, 160–63
Brenner, Sydney, 76–77, 82, 90, 106; embryology and, 144; genetic code work by, 97–98, 122; mRNA work by, 117–20; neurobiological interests of, 151; paper on triplet code, 135; phage genetics work by, 132, 135–36
Bressler, 130
British Broadcasting Corporation (BBC), 84, 86
Bromotyrosine, 91
Bronowski, Bruno, 145
Brooklyn Polytechnic, 94
Buchner, Edward, 32

Caenorhabditus elegans, 144
Caius College, 40
California, 14, 144–45
California Institute of Technology, 58, 61, 73, 120, 126
Callender, Jane, 84
Cambridge University, 4, 13–14, 19, 21, 39–40, 43–44, 63–64, 67, 77–78, 81, 87, 96, 103, 118, 130–31, 148, 158; *see also specific colleges*
Carbon, 54, 56, 58, 61, 165–67
Carboxyl (CO) groups, 56
Carnegie-Mellon University, 159
Cartilage, 67
Catholic Church, 94
Cavendish, Henry, 39
Cavendish Laboratory, 4, 23, 38–41, 44–45, 48, 59, 63, 65, 103
Central dogma, 109, 168
Cerebral cortex, 156, 157
Chadwick, James, 39–40
Chain terminators, 133–34
Chargaff, Erwin, 36, 77, 80, 85, 164
Chargaff's rules, 64–65, 86, 164
Chemical and Engineering News, 18
Chemistry, 12, 17, 61, 75, 93, 148; of DNA, 38; inorganic, 12; lack of knowledge of on part of many physicists, 18; of macromolecules, 32, 61; Nobel Prize for, 43, 51; organic, 12, 17; of polypeptide synthesis, 54; protein, 107; quantum mechanics application to, 58
Chick: fibroblasts, phagocytosis of magnetic ore by, 22; lysozyme, 104–5

Index

Chromatography, 92; protein fingerprinting and, 105; tRNA and, 113
Chromosomes: approximate size of, 5; DNA in, 144; occurrence of cell genes on, 33, 112; selfish DNA in, 31, 147
Churchland, Pat, 153–54
Churchland, Paul, 153
Church of England, 8
Circular DNA, 72
Cis-trans test, 126
Cistrons, 98, 126
Cochran, William, 59
Codons, 90, 93, 97, 108, 110, 128–32, 134, 139, 169
Cognitive science, 149, 157
Cohn, Mel, 145
Coil of Life, The (Moore), 100
Cold Spring Harbor: annual meeting (1966), 143; Symposium (1953), 66, 76
Co-linearity, 102
Colinvaux, Raoul, 13
Collagen, 67
Collingwood, Edward, 16
Color vision, 18, 158
Combination code, 96–97
Comma-free code, 99–101, 139
Commoner, Barry, 77
Complexity: biological, 138; species, 24
Condensation, 161
Cones, number of in eye, 18
Congregational Church, 10
Consciousness: experimental study of, 157–58; mystery of, 18
Conservation laws, 138
Continuous creation theory, 94
Cosmology, 71, 91, 99, 148
Coulsen, Charles, 59
Cowan, Max, 153
Cowan, Pauline, 68
Cozzarelli, Nick, 72
Creationism, 24–25, 30
Crick, Anne Elizabeth (née Wilkins), 7–10
Crick, Arthur, 13
Crick, Doreen (née Dodd), 8, 14, 43
Crick, Francis: adaptor hypothesis and, 96; α helix and, 57, 59, 68; Biochemical Congress at Moscow (1961) and, 130–31; biological training of, 6; biophysics and, 13; birth, 7; botanical interests of, 9; brain visual system and, 150–58; Cambridge home of, 44; chemistry training and, 9, 12, 17; childhood and early years of, 3, 7–14; collagen work by, 67; directed panspermia and, 148; divorce of, 43; DNA double helix co-discovery by, 4, 36; DNA structure work by, 60, 64–79, 90; early molecular biology interests of, 17–18, 20, 22–23; early pure scientific research of, 16–17, 19, 20–23; early scientific interests of, 8–11; early work experience of, 13–16, 43; education of, 3–4, 12–13, 42; embryology and, 144; error-correcting device in DNA replication and, 111; experiences with books and movies about DNA, 81–88; extraterrestrial intelligence interests of, 148; family life of, 129; feelings on discovery of double helix, 78–79; first crystallographic talk of, 49; first marriage of, 8, 14, 43; first published papers of, 22; first research seminar of, 49; genetic code work, 89–101, 122; "gossip test" theory of, 15–23; graduate school and, 4, 51, 64–65, 67, 71; impressions of *The Double Helix,* 80–81; later years of, 143–63; *Life Itself,* 148; mathematics and, 9, 12, 16; mRNA work by, 116–21; neurobiological interests of, 17–18, 146–63, 162; Nobel Prize awarded to, 52; paper on comma-free code, 100; papers on DNA structure, 66–67, 71, 73, 76–77, 91; paper on genetic code, 95; paper to Society of Experimental Biology, 108; paper on triplet code, 135; phage genetics work by, 124–36; Ph.D. thesis of, 64, 67; physics training and, 6, 12–13, 16; protein fingerprinting research by, 102–7; protein structure research by, 23; religious views of, 3, 10–11, 17, 109; RNA Tie Club and, 95; *Scientific American* article by, 71; scientific intelligence experience of, 14; on scientific research as an activity, 137–42; second marriage of, 8, 20–21, 43–44, 64, 77, 81–82, 86, 98, 105, 127, 129, 134, 145; "selfish DNA" work by, 147; on sound biological theory, 137–42; sports interests of, 9–10, 12, 14; suppressors and, 126; Tangier meeting and, 130; weapons research during World War II, 13, 15; X-ray crystallography work and, 40–52
Crick, Harry, 7, 9–10, 12

Crick, Jacqueline, 105
Crick, Michael, 8, 14, 44
Crick, Odile, 8, 20–21, 43–44, 64, 77, 81–82, 86, 98, 105, 127, 129, 134, 145
Crick, Tony, 8–10, 14
Crick, Walter, 7
Crosses, gene, 127, 132
Cross-feeding, 27
Cross-overs, 123
C2 space group, 66
Cysteine, 92, 170
Cystine, 92
Cytoplasm, 103, 170; physical properties of, 22; and ribosomal protein synthesis, 116–17, 146; RNA export to, 117; RNA in, 89, 110, 120, 140
Cytosine, 62, 65, 71, 90, 97, 99–100, 132, 164–66, 169–70

Darwin, Charles: blending inheritance and, 27; invalidation of creationism by, 25; Linnean society and, 25; natural selection theory of, 25, 28
da Vinci, Leonardo, 135
Dawkins, Richard, 28–30, 147
de Hoffmann, Frederic, 145
Delbrück, Max, 61, 64, 76, 95, 126
Deletions, 123–25, 132, 135; overlapping, 123–24
Demerec, Milislav, 77
Deoxyribonucleic acid, *see* DNA
Deoxyribose, 164–65
Diaminopimelic acid, 91
Dickerson, Richard, 73
Dihedral angle, 55–56
Dinosaurs, extinction of, 24
Diploidy, 106
Diversity, of species, 24
DNA: A form, 85, 112; α helix, 53–61, 68; backbone of, 164, 166; bacterial cell, 168; β helix, 53–55; B form, 64, 67, 85; books and movies about, 81–88; chemistry of, 38; chromosomal, 144; circular, 72; coding for proteins, 93; composition of genes, 22; control sequences for gene induction and repression, 167; C2 space group, 66; double helix, 3, 36, 62–82, 89–90, 102, 111, 116, 122, 125, 165; ease of understanding concept of, 63; error-correcting device in replication of, 111; fibers, 52, 64, 66, 68, 72–73, 112; four-letter alphabet of, 36, 169–70; impurities in, 37; left-handed structure, 73; length of presence on Earth, 62; linking number and, 72; linking of unlinked circles of, 72; mitochondrial, 170; molecule length, 36; molecule shape, 164; molecule size, 33; mutation rate per effective base pair and, 27; nonsense sequences, 146; nuclear, 116, 140; as nucleoprotein constituent, 33; origin of, 138; parasitic, 147; phage, 119; phosphate groups, 164; recombinant, 102, 146; recombinant techniques in *Drosophila* genetics, 25; radiation and, 22, 111; replication, 27, 62, 66, 77, 90, 111, 166; selfish, 31, 147; sequence complementarity of, 63; sequencing, 146; side-by-side structure theory of, 71–72; single-stranded, 168; structural similarity to RNA, 165; structure of, 60, 64, 67, 91, 100, 108; sugar groups, 164; as template for amino acid side chains, 95; transforming factor of pneumococcus and, 36–38; two chains of, 77, 165; unlinking of linked circles of, 72; viral, 61, 164; Z-, 73; *see also* Base pairs; Nucleotides
Dominance: allelic, 26; incomplete, 26
Donohue, Jerry, 65, 85
Doty, Paul, 63
Double Helix (movie), 84
Double Helix, The (Watson), 80
Dreams, theory of, 161–62
Dreyer, William, 135
Drosophila, 25, 147, 151
Dulbecco, Renato, 146
Dyadic symmetry, 65

Earth, true age of, 11, 148
Eccles, Sir John, 154
E. coli: bacteriophage T4 infection of, 119, 122–24; introns and, 147
Egg, sexual reproduction and, 31
Egg white lysozyme, 104
Eigen, Manfred, 111
Eigenvalues, 13
Eighth Day of Creation, The (Judson), 81
Electron: density, 43, 48; density mapping, 47; discovery of, 39; microscope, 20

Index

Electrostatic interactions, 61
Embryology, 30–31, 140, 163
Encounter, 146
End chains, 90; STOP codon and, 169–70
Energy: of attraction, 140; free, 27; of repulsion, 140
Environment, role of in natural selection, 30
Enzymes, 32–33, 35, 60; induction of, 118; pancreatic production of, 23
Ephrussi, Boris, 103
Error-correcting device, 111
European Conference on Visual Perception, Seventh, (Cambridge), 163
Evolution, 31; α helix and, 59; difficulty in studying, 30–31; hostility of creationists toward, 25; natural selection and, 5, 111; process of, 138–39; of stars, 137; over time, 25–26, 29; as tinkerer, 156
Experimental Cell Research, 22
Extinction, of species, 24
Eyes, photoreceptors of, 155

Fell, Honor B., 21, 68
Fermentation, 32
Feynman, Richard, 126
Fibroblasts, chick, 22
Fibrous proteins, 35
Fisher, R. A., 26, 107
Fivefold symmetry, 45; α helix and, 57
Fleming, Alexander, 104
Fodor, Jerry, 160
Formylmethionine, 92
Fossil record, 11
Fouracre, Ronnie, 81
Fourier Transforms, 47, 59, 112
Fractionation methods, 113
Franklin, Rosalind, 71; and A form DNA, 112; and B form DNA, 67; movie characterization of, 84–86, 88; paper on DNA structure, 66; personal background of, 68–69, 86–87; and Tobacco Mosaic Virus, 75; X-ray crystallography work on DNA structure, 52, 64–65, 68–69
Free energy, 27
Freese, Ernst, 125
Freud, Sigmund, 161
Freudenthal, H., 101
Frog's eye, insect detectors in, 148
Fundamental particles, 137

Galactose, 118
Gamow, George, 71, 91–96
General Chemistry (Pauling), 12
Gene(s): and amino acid sequence in proteins, 36, 54, 89; bacterial transformation and, 37; chemical structure of, 74, 112; control of bacteria, 143; control of protein synthesis, 102; cross-over, 123; crosses, 132; cytoplasmic, 103; deletions, 135; DNA between, 147; DNA composition of, 22, 37–38; frequency of recombination, 122; hybrid, 123; induction, 167; isolation, 162; leaky mutants, 125; magnification, 106; mapping, 122–24, 126, 128, 147; molecular structure of, 22; mutants, 26, 102–4, 125, 128–29, 131; nature of, 33, 35, 64, 143; nonleaky mutants, 125, 128–29; nuclear, 103; occurrence of on chromosomes, 112; omissions, 135; polysaccharide coats of, 38; products, 128, 136; protein component of, 38; protein synthesis control by, 22, 33; r_{II}, 122, 126, 131; random mutation, 26; replication, 33, 35, 87, 114–15, 143; repression, 167; RNA intermediate in protein synthesis and, 22; size of, 31–32; structure, 107, 114; suppressors, 125; three-dimensional structure of, 35; transfer of bacterial, 118; tRNA, 147; wild-type, 122; as units of instruction, 31; *see also* Genetic code; Genetics; Mutations
Genetic code, 76, 78, 89–101, 108–9, 117, 120, 122, 125, 136, 143, 169–70; combination code, 96–97; comma-free code, 99–100, 139; elucidation of, 3–4, 81; triplet code, 133, 135; uniformity of, 90, 148; universality of, 97; *see also* Gene(s); Genetics; Mutations
Genetic disease, 105–6
Genetics: bacterial, 36–37, 64, 118; behavioral, 151; classical, 65, 150; control mechanisms, 31; crosses, 127; general ignorance toward, 25; Mendelian, 12, 25–26, 28, 103, 106, 138; molecular basis of, 25; phage, 122–36; recombination, 122–23; of speciation, 24–38; variation, 31; *see also* Gene(s); Genetic code; Mutations
Genotype, 106
Geophysics, 137
Glasgow University, 59

Globular proteins, 35
Glutamate, NMDA receptor for, 163
Glutamic acid, 105, 170
Glutamine, 91, 170
Glycine, 54, 167, 170
Gold, Thomas, 94
Goldblum, Jeff, 84, 88
"Golden Helix, The," 78, 81, 106, 120
Golomb, Sol, 101
Gosling, Raymond, 66, 67, 88
Gradients, 144
Green College, 84
Griffith, John, 85, 86, 99, 100
Group theory, 13
Guanine, 62, 65, 71, 90, 93, 97, 99–100, 128–29, 132, 134, 164–66, 169–70

Haldane, J. B. S., 24, 106
Hamlet (Shakespeare), 28
Handedness, of biological molecules, 45
Haploidy, 106, 111
Hardy Club, 78, 148
Hartridge, Hamilton, 18–19
Harvard Medical School, 96
Hauptman, Herbert, 43
Hayes, William, 135
Helix: α, 53–61, 68; β, 53–55; double, 3, 36, 62–82, 89–90, 102, 111, 116, 122, 125, 165; integer axes of, 57–58; molecular simplicity of, 57–59; noninteger screw of, 59, Pauling's α, 64; peptide bonds planar in, 58; screw axis of, 57–58
Helmholtz Club, 154
Hemoglobin, 47; heavy atoms added to, 103; horse, 41, 48; human, 105–6; structure of, 41, 51; three-dimensional structure of, 38; X-ray diffraction studies of, 41
Hershey, Al, 64
Hill, A. V., 19, 21
Hinshelwood, Sir Cyril, 18, 77
Hippocampus, 157
Histones, 70, 144
Hoagland, Mahlon, 96
Hodgkin, Alan, 78, 148
Holley, Robert, 113
Homopolar bonds, 61
Hopkins, Gowland, 40
Hotchkiss, Rollin, 37
Howard, Alan, 84
Hoyle, Frederick, 94, 99
Hubel, David, 146
Hughes, Arthur, 21–22
Huxley, Andrew, 148
Hydrodynamics, 13, 16, 22
Hydrogen, 54, 65, 167
Hydrogen bond, 18, 55–56, 62, 65, 71, 95, 115, 165
Hydroxyproline, 91

Induction, gene, 167
Infection, bacterial, 104
Ingram, Vernon, 103, 105
Inheritance: blending, 26–27; genetic basis of, 36–37; Mendelian, 103; particulate, 26–27
Insects, gradients in cuticle of, 144
Institut Pasteur, 117
Insulin, 92–93
Integer axes, α helix and, 57
Integer screw, α helix and, 58
International Congress of Biochemistry, First, 43
Interpeak distances, in electron density map, 47
Introns, 146–47, 168
Ions: exchange columns, 104; movement of in giant axon of squid, 21; sodium, 65
Isomorphous replacement, 49–51, 73
Isotope separation, 19
Itano, Harvey, 105

Jackson, Mick, 84, 88
Jacob, François, 5, 117–19, 120
Jenkin, Fleeming, 27
Judson, Horace Freeland, 81, 118

Karle, Jerome, 43
Keller, Walter, 72
Kendrew, John, 23, 50–51, 56, 65, 81; α helix and, 57–58; myoglobin and, 67
Keratin, X-ray diagrams of, 53
Keynes, Richard, 21
Khorana, Gobind, 101
Kieckhefer Foundation, 145
King's College, London, 19–20, 66, 68, 75, 78, 85, 118

Index

Klug, Aaron, 86, 144
Kornberg, Arthur, 77, 111
Kornberg, Roger, 144
Kornberg enzyme, 111
Kreisel, Georg, 16, 148

La Jolla, California, 144, 153
Lattice, crystal, 46
Lawrence, Peter, 144
Lea, D. A., 21
Lederberg, Joshua, 37
Lennox, Edward, 145
Lerman, Leonard, 135
Leucine, 110, 170
LeVay, Simon, 153
Levine, Phoebus, 33
Life Itself (Crick), 148
Life Story (movie), 4, 84–86, 88
Light: microscopy, 20, 41; ultraviolet, 20; wavelength of visible, 20
Linguistics, modern, 152
Linking number, 72
Linnean Society, 25
Lipmann, Fritz, 77
Locke, Michael, 144
London, 14, 39, 43, 53, 85, 145
Luria, Salva, 64
Luzzati, Vittorio, 85–87
Lysenko, 130
Lysozymes: chick, 104–5; egg white, 104; guinea fowl, 104; phage, 136; tear drop, 104–5

McCarty, Maclyn, 36–37
McClelland, Jay, 159
MacLeod, Colin, 36–37
MacLeod, Donald, 153
Macroevolution, 29–30
Macromolecules, 31–32, 40, 61
Maddox, John, 16
Magnetic ore, phagocytosis of by chick fibroblasts, 22
Magnetism, 16, 22
Magnetron, 19
Magnification, genetic, 106
Mandell, George, 157
Mapping, gene, 122–24, 126, 128, 147
Mark, Herman, 40
Markham, Roy, 21, 120
Marr, David, 158
Massachusetts Institute of Technology, 73, 153, 158
Massey, H. S. W., 15, 19
Mathematics, 12, 16, 160
Maxwell, James Clerk, 39
Maxwell's equations, 39
Medawar, Peter, 76
Medical Research Council (MRC), 18–19, 21, 23, 43, 87, 145
Medicine, Nobel Prize for, 52, 146
Mee, Arthur, 8
Mellanby, Sir Edward, 19, 21–22
Memory, 158, 161
Mendel, Gregor, 25; *see also* Genetics
Mercury, 49–50
Meselson, Matt, 120, 130
Metabolism, cellular, 33
Methionine, 92, 170
Methyl (CH_3) groups, 54
Microevolution, 25–27, 30–31
Microscopy, 20
Mind into Matter (Delbrück), 61
Mines, noncontact, 15
Minton, John, 78
Mirsky, Alfred, 37
Mr. Tompkins Explores the Atom (Gamow), 92
Mitchell, Peter, 86
Mitchison, Graeme, 144, 161
Mitchison, Murdoch, 114
Mitochondria, 103, 170
Models: Astbury's keratin, 53; behavioral, 149; Bragg's, Kendrew's, and Perutz's α helix, 55–58; brain, 115; of chemical molecules, 63; of DNA structure, 64, 73; giant axon potential in squid, 148; good biological, 115; neural net, 159–60; Pauling's α helix, 58, 64; Penroses's gene replication, 114–15; polypeptide backbone, 59; side-by-side DNA structure, 72; three-chain DNA, 85; Watson and Crick's DNA structure, 65, 73–74, 85, 114–15
Molecular biology, 17, 22, 61, 162–63; and borderline between living and nonliving, 17–18, 20; brief outline of classical, 3, 5, 164–68; chemical structure of gene as central problem of, 74; theory in, 107–15
Molecular psychology, 163
Molecules: adaptor, 96; approximate size of, 5; chemical, 63; handedness of, 63;

Molecules *(continued)*
King's College biophysics research and, 20; as parasites, 147; tRNA, 113
Molteno Institute, 96
Monkeys, visual system in, 150–53, 156, 158
Monoclonal antibodies, 146
Monod, Jacques, 109, 117–18, 145, 149
Monomers, 34; *see also specific monomers*
Montefiore, Hugh, 29–30
Moore, Ruth, 100
Morse code, 90, 102, 139
Mott, Professor Neville, 63
Muller, Hermann, 74, 106
Muscle, biophysics of, 19
Mutations: acridine, 135–36; amino acids and, 126; amino acid string alterations, 136; beneficial, 26; deleterious, 26; deletions, 123, 132; distributions, 136; double, 129, 133; gene, 102–4, 123–36; genetic diversity and, 26–27; leaky, 125; multiple, 136; nonleaky, 125, 128–29; nonsense, 131; phenotype, 132; point, 123; proflavin, 125; quadruple, 136; random, 26, 35, 147; rate, 111; during replication, 29; revertant, 126; single, 136; suppressors, 125–26, 128–29; triple, 132–33, 136; viral, 133; wild-type, 126, 129, 131–32, 134–35; *see also* Gene(s), Genetic code, Genetics
Mutons, 98
Myoglobin, 68; heavy atoms added to, 103; structure of, 51; three-dimensional structure of, 38

Natural selection: appearance of planned design due to, 25; biological replication and, 5; brain as product of, 161; contemporary acceptance of, 30; contemporary understanding of, 31; as cumulative process, 5, 28–29; fair criticisms of, 30–31; evolution and, 111; genetic diversity and, 26–28; power of, 29; randomness of, 30; rate of, 30; remarkable structures built by, 61; role of environment in, 30; selective pressures and, 30; selfish DNA and, 147; sex and, 31; theory of, 115; unplanned nature of, 28
Natural Theology (Paley), 25
Nature, 16, 66, 71–72, 77, 91, 134
Nauta, Walle, 149
Nematodes, 144, 151
Neocortex, 159
Nerve cells, 149
NetTalk, 159
Neural nets, 158–61
Neuroanatomy, 149–52
Neurobiology, 17–18, 51, 107, 149, 159, 162
Neurons, 150, 154–60, 162
Neurophilosophy (Churchland), 154
Neurophysiology, 149–51, 156–57
Neurospora crassa, 33
Neutron, discovery of, 40
Newtonian mechanics, 138
Nicholson, William, 84, 87
Nirenberg, Marshall, 100, 130
Nitrogen, 54, 165–67; atomic, 59
NMDA, receptor for glutamate, 163
Nobel Prize, 18, 19, 32, 33, 40, 43, 51, 52, 94, 146
Noninteger screw, α helix and, 59
Nucleic acids: molecular structures of, 52; tetranucleotide hypothesis and, 33; ultraviolet light absorption by, 21; *see also* DNA; RNA
Nucleoproteins, 33
Nucleosomes, 144
Nucleotides, 62, 70, 135, 140, 164; acridines between, 135; DNA, 89–90; looped out, 125; number of in human cell, 168; order of, 34, 164; RNA, 89–90, 119; sequences of, 73, 108–9, 128–29; tautomeric nature of, 111; triplets, 122–36, 139, 169; *see also* Base pairs; *specific nucleotides*
Nucleus, cellular, 89–90, 103, 116

Occam's razor, 138
Occlusion, 155
"Ode on a Grecian Urn" (Keats), 50
Olby, Robert, 81
Omissions, 135
"On Degenerate Templates and the Adaptor Hypothesis" (Crick), 95
One gene–one enzyme hypothesis, 33
Ontogeny, 31
Orgel, Leslie, 95, 126; error-correcting device in DNA replication and, 111; extraterrestrial intelligence interests of,

Index

148; genetic code work of, 99, 134–35, 147; paper on comma-free code, 100; paper on triplet code, 135
Origin of Species, The (Darwin), 25
Overlapping, *see* Triplets
Oxford University, 58, 77, 84; *see also specific colleges*
Oxygen, 41, 54, 56, 165–67

PaJaMo experiment, 117–19
Paley, William, 25, 28
Pancreas, function of, 23
Panspermia, directed, 148
Parallel distributed processing (PDP), 159–60
Parallel Distributed Processing (Rumelhart and McClelland), 159
Parasitic DNA, 147
Pardee, Arthur, 117, 119
Paris, 85–86, 98, 103, 117, 145
Particulate inheritance, 26–27
Path to the Double Helix, The (Olby), 81
Patterson, Lindo, 47–48
Patterson density map, 47–49
Pauling, Linus, 12, 65, 69, 85; and α helix, 58–60, 64, 75; and hemoglobin, 105; hydrogen bond and, 18; importance of to molecular biology, 60–61; and sickle-cell anemia, 60, 105; and vitamin C, 58
Pauling, Peter, 85, 88
Penicillin, 17, 104
Penrose, Lionel, 114–15
Penrose, Roger, 114
Peptide bonds, planar in α helix, 58
Perception: scientific study of, 154; utilitarian theory of, 156
Periodic table, 101
Perutz, Max, 23, 51, 64–65, 81–82, 85; and α helix, 57–59; and hemoglobin, 105; movie characterization of, 88; and protein structure, 39–48
Peterhouse College, 78
Phage Group, 61, 64, 75
Phages, 64, 124, 147; DNA, 119; genetics, 122, 124–25, 127–28, 130–36; lysozyme, 136; *see also* Bacteriophages
Phagocytosis, of magnetic ore by chick fibroblasts, 22
Phase sequence, 128–29
Phase shift, 100, 131
Phenotype, 107, 132
Phenylalanine, 130, 170
Philosophical Transactions of the Royal Society, 134–35
Phosphate groups, 70, 164
Photoreceptors, 155
Physical chemistry, 75; α helix and, 59
Physics, 12–13, 16; difference of from biology, 5–6, 137–39; entry of into biology, 19
Physiology, 18, 21; Nobel Prize for, 52
Pigott-Smith, Tim, 84
Plane groups, 45
Plants: extant species, 24; viruses, 21
Plaques: bacterial, 122, 127, 133; mutants, 123; picking of, 127; wild-type, 123
Pneumococcus, transforming factor of, 36–38
Poggio, Tomaso, 158
Pohl, William, 72–73
Poisson distributions, 136
Polymers, 34, 36, 40, 91, 166; *see also specific polymers*
Polypeptide chains, 34, 146; backbone of, 34, 54–55, 59; basic chemical formula for, 166; chemistry of synthesis, 54; cutting of by trypsin, 105; folding, 53, 55; NH group donors, 56; nonsense sites, 131; repeats, 53–54; side-chains, 34, 54–55, 166; synthesis, 36, 117, 166–67; *see also* Polypeptides; Proteins
Polypeptides, 57, 59; *see also* Polypeptide chains; Proteins
Polyphenylalanine, 100, 130
Polysaccharide coat, of gene, 38
Princeton University, 153
Probability of God, The (Montefiore), 29–30
Proceedings of the National Academy of Sciences, 58, 100
Proceedings of the Royal Society, 41, 57
Proflavin, 125, 128
Protamines, 70
Proteins: amino acid residues in, 109; amino acid sequences in, 34, 36, 54, 89, 91–92, 103; collagen, 67; denatured, 35; distinct types of, 61; DNA coding for, 93, 166; enzymes as, 32–33; fibrous, 35; fingerprinting, 102–7; folding, 3, 53, 139–40, 167; function dependent on three-dimensional structure, 35; gene determination of, 33; as genetic material in

Proteins *(continued)*
transforming factor, 37; globular, 35; as nucleoprotein constituent, 33; RNA coding for, 93; shape, 34–35; sizes, 167; structure, 23, 41, 45–49, 51, 60, 126; symmetry elements, 45; synthesis, 21–22, 33, 54, 87–91, 97, 102, 110, 116–17, 119, 120, 130, 143–44, 167; three-dimensional structure, 4, 34, 36, 38, 40–41, 49, 55; X-ray diffraction patterns of, 4, 35, 64; *see also specific proteins*
Psychology, 157, 160–62; behaviorism and, 149; molecular, 163; parallel distributed processing and, 159–60; of vision, 18
Psychophysics, 149, 157; of vision, 153
Purines, 65, 73, 125, 164
Putnam, Frank, 81
Pyrimidines, 65, 73, 125, 164

Quantum electrodynamics, 13, 138
Quantum mechanics, 13, 58, 61, 63, 115
Quarks, 13

r_{II} gene, 122, 126, 131
Radar, military applications of, 19
Ramachandran, V. S., 153–54, 156
Randall, John, 19, 64, 69, 85
Reagan, Ronald, 25
Recombinant DNA, 102, 122, 146
Recons, 98
Red blood cells, 41, 106
Reductionism, 154
Reflection, 45, 155
Relativity, 63, 138
Religion, incompatibility with some scientific beliefs, 11
REM sleep, 161
Replication: cellular, 27; DNA, 62, 77, 90, 111, 166; error rate per step of, 111; exact, 27; gene, 33, 35, 87, 114–15, 143; geometrical nature of, 27, 137; mutation and, 27, 29; RNA, 89; semiconservative model of, 27; template concept of, 27
Repression, gene, 167
Retroviruses, 146, 168
Revertants, 126
Ribonucleic acid, *see* RNA
Ribose, 165
Ribosomes: gene production of after transfer, 118; protein synthesis on, 144, 146, 167; RNA in, 116–17, 119–20, 140
Rich, Alex, 4; Biochemical Congress at Moscow (1961) and, 131; collagen work by, 67; DNA structure work by, 73; genetic code work by, 94, 96; RNA Tie Club and, 95
Riley, Monica, 119
RNA: bacterial cell, 117; base pairs, 27; bases, 89–90, 119; base sequences, 102; coding for proteins, 93; cytoplasmic, 89, 110, 120, 140; export to cytoplasm, 117; four-letter language of, 117, 169–70; fractionation of tRNA, 113; intermediate in protein synthesis, 22; length of presence on Earth of, 62; messenger (mRNA), 89, 106, 116–21, 140, 143, 146, 167; mutation rate per effective base pair, 27; as nucleoprotein constituent, 33; replication, 27, 89; retroviruses, 168; ribosomal (rRNA), 116–20, 140; shape, 164; single-stranded, 117, 167; sizes of, 117; structure, 165; synthesis, 110; as template, 95–96, 110; transfer (tRNA), 96, 113, 117, 147, 167–68; viruses, 146, 164; Volkin-Astrachan, 119
RNA Tie Club, 95–96, 100
Rock crystals, 138
Rockefeller Institute, 36–37, 76–77
Rosenberg, Charles, 159
Rotation: angles of, 55; axes, 45
Rothschild, Lord Victor, 78
Royal Institution, 53, 63
Royal Society, 18, 78, 114, 135
Rumelhart, David, 153, 159
Russian Atomic Energy Research establishment, 130
Rutherford, Ernest, 39; and α helix, 58

Salk, Jonas, 145
Salk Institute for Biological Studies, 144–46, 149, 153, 158–59
Sanger, Frederick, 34, 93, 103, 105–6
Schroedinger, Erwin, 18
Scientific American, 13, 71, 114
Screw axis, α helix, 57–58
Search, The (Snow), 21
Sejnowski, Terry, 159

Index

Selfish DNA, 147
Selfish Gene, The (Dawkins), 147
Semiconservative model of DNA replication, 27
Sequence hypothesis, 108
Sequences: amino acid, 93, 97, 102–3, 108–9; base, 128–29; complementary, 63; DNA, 62, 74, 146; phase of, 128–29
Sexual reproduction, natural selection and, 31
Shape-from-shading, 155
Shaw, Gordon, 154
Shrinkage states, 46
Sickle-cell anemia, as molecular disease, 60, 105–6
Side-by-side hypothesis of DNA structure, 71–72
Side-chains, 34, 36, 54–55, 95, 166; charges on, 54; DNA and RNA as templates for, 96; hydrogen atom, 54; methyl group, 54; total number known, 166
Silicon, 41
Silicon chips, 160
Slow viruses, 164
Snow, C. P., 15, 21
Society for Experimental Biology (London), 108
Sodium chloride, 41
Sodium ion, 65
Space groups, 45
Spatial repeat unit, 45
Speciation, genetics of, 24–38
Sperm, 31
Splicing, gene, 146
Squid, ion movement in giant axon of, 21, 148
Stanford University, 159
Stars, 138; evolution of, 137
Stereopsis, 155
Stent, Gunther, 76, 80
Stevenson, Juliet, 88
Stokes, A. R., 66
STOP codon, 169–70
Strangeways Laboratory, 4, 21–22
Streisinger, George, 135–36
Sugar, 32, 164
Sumner, James, 32–33
Suppressors: FC series, 128–29; internal, 126; mutants, 125–26, 128–29; P series, 128
Sutherland, Stuart, 160
Symmetry, 45, 57, 59, 65
Synge, Dick, 92
Szent-Györgyi, Albert, 94–95
Szent-Györgyi, Andrew, 95
Szilard, Leo, 149

Tamm, Igor, 130
Tatum, Edward, 33; bacterial genetics work by, 37
Tautomeric forms, 65, 75, 85, 111
Tears, human, lysozyme in, 104–5
Temin, Howard, 146
Template concept of DNA replication, 27, 35
Tendons, 67
Tetranucleotide hypothesis, 33–34, 36
Theology, 158
Thermal energy, 55
Thermal motion, 57, 115
Thom, René, 136
Thomson, J. J., 39
Thymine, 62, 65, 90, 97, 99–100, 128–29, 132, 164–66, 169–70
Tissue culture techniques, 21, 68
Tobacco Mosaic Virus (TMV), 75
Topoisomerase II, 72
Topology, 136
Transitions, 125
Transversions, 125
Triplet code, 133, 135
Triplets, nucleotide, 93, 97, 99, 101, 122–36, 139, 169; chain terminator, 133–34; nonoverlapping, 99, 131; nonsense, 99; overlapping, 97–98; sense, 99
Trypsin, 105
Tryptophan, 54, 91, 170

Ultraviolet microscopy, 20
"Uncles and aunts" (r_{II} gene mutants), 131
Unit cell, 45, 50, 65
Universe, age of, 148
University College, London, 12–14, 19, 114
University of California: Berkeley, 19, 119; Irvine, 153–54; San Diego, 153–54, 157
University of Vienna, 40
Uracil, 90, 100, 130, 134, 165, 169–70
Urease, 33

Valine, 105, 169–70
Vand, Vladimir, 59
van der Waal's forces, 61
Variation: genetic, 31; preservation of, 26
Villa Serbelloni, 136
Virology, 120
Viruses: AIDS, 146, 168; bacteriophages, 37, 61, 64, 107; DNA, 164; flu, 168; mating of, 124; mutants, 122, 133; plant, 21; polio, 45, 168; RNA, 146, 164; size of, 20; slow, 164; spherical, 45; structure of, 4
Viscosity, of water, 13
Vision: color, 158; psychology of, 18; psychophysics of, 153
Vitamin C, 58, 94
Volkin, Elliot, 119
Volkin-Astrachan RNA, 119

Waddington, Conrad, 136
Walden, Leonard, 13
Wallace, Alfred, 25
Wang, James, 72–73
Water, 13, 55–56, 65
Watson, Elizabeth, 88
Watson, James, 82; α helix work by, 59–60, 68; bacterial genetics and, 64; birdwatching interests of, 46; Cold Spring Harbor Symposium talk at, 76; DNA double helix discovery and, 4, 36; DNA structure and, 60–79, 90; *The Double Helix,* 80; experiences with books and movies about DNA, 81–88; feelings on discovery of double helix, 79; genetic code work by, 89–101; mRNA work by, 116; nature of genes and, 64; Nobel Prize awarded to, 52, 81; papers on DNA structure, 66–67, 71, 76–77, 91; Phage Group and, 61; protein synthesis and, 90; RNA Tie Club and, 95; viral structure and, 4; X-ray crystallography and, 45
Wave mechanics, 13
Welch, Lloyd, 101
What Is Life? (Schroedinger), 18
Whose Life is It, Anyway? (movie), 82
Wiesel, Torsten, 146
Wilcox, Michael, 144
Wilkins, Ethel, 7–8
Wilkins, Maurice, 19, 68; DNA structure work by, 60, 75; personal interests of, 20–21; movie characterization of, 84–86, 88; Nobel Prize awarded to, 52; paper on DNA structure, 66; ultraviolet microscopy by, 20; X-ray crystallography work on DNA fibers, 64–65, 68–69
Wilson, H. R., 66
Wood's Hole, 94
World War I, 7
World War II, 3, 13–15, 40, 64; rise in influence of science due to, 19; weapons research during, 13, 15

X-ray crystallography, 4, 21, 41–44, 46–49, 50, 75, 103; of A form DNA, 112; of α helix, 57, 59; of B form DNA, 67; Bragg's law for, 40; diffraction pattern, 45; of DNA fibers, 64, 68, 72–73; of DNA structure, 65, 67, 71; of hemoglobin, 41; major problem of, 42; of myoglobin, 68; of proteins, 41, 51–53, 55, 64; of synthetic peptides, 59; theory of, 45; of three-dimensional structure of proteins, 35, 40–42; of Z-DNA, 73
X-ray spots, 42–43, 48, 50, 57, 73, 112

Ycas, Martynas, 94–96
Yeast, 24, 32, 147

Z-DNA, 73
Zeeman, Christopher, 136
Zipser, David, 157